Firefighter Fatalities in the United States in 2002

U.S. Department of Homeland Security
Federal Emergency Management Agency
U.S. Fire Administration

July 2003

In memory of all firefighters who
answered their last call in 2002

To their families and friends

To their service and sacrifice

 **U.S. Fire Administration
Mission Statement**

As an entity of the Federal Emergency Management Agency, the mission of the United States Fire Administration is to reduce life and economic losses due to fire and related emergencies, through leadership, advocacy, coordination, and support. We serve the Nation independently, in coordination with other Federal agencies, and in partnership with fire protection and emergency service communities. With a commitment to excellence, we provide public education, training, technology, and data initiatives.

On March 1, 2003, FEMA became part of the U.S. Department of Homeland Security. FEMA's continuing mission within the new department is to lead the effort to prepare the nation for all hazards and effectively manage federal response and recovery efforts following any national incident. FEMA also initiates proactive mitigation activities, trains first responders, and manages Citizen Corps, the National Flood Insurance Program and the U.S. Fire Administration.

ACKNOWLEDGEMENTS

This study of firefighter fatalities would not have been possible without the cooperation and assistance of many members of the fire service across the United States. Members of individual fire departments, chief fire officers, the National Interagency Fire Center, United States Forest Service personnel, the United States military, the Department of Justice, NFPA International, and many others contributed important information for this report.

IOCAD Emergency Services Group of Emmitsburg, Maryland (a division of IOCAD Engineering Services, Inc.) conducted this analysis for the United States Fire Administration (USFA) under contract EME-98-CO-0202-T017.

The ultimate objective of this effort is to reduce the number of firefighter deaths through an increased awareness and understanding of their causes and how they can be prevented. Firefighting, rescue, and other types of emergency operations are essential activities in an inherently dangerous profession, and unfortunate tragedies do occur. This is the risk all firefighters accept every time they respond to an emergency incident. However, the risk can be greatly reduced through efforts to increase firefighter health and safety.

Photographic Acknowledgments

The USFA would like to extend its thanks to the following individuals for providing photographs for this report:

John Severson, Indianapolis Star

Patrick Schneider, Charlotte Observer

J. Intintoli, Daily News Journal

August L. Meyland III, News-Record

Firefighters console one another in front of the wreckage of the tanker which was operated by Firefighter Cassandra "Sandy" Myers Billings Powell at the time of her death. The apparatus crashed onto it's roof after leaving the roadway 21

Derek Neas, Duluth News Tribune

The photo depicts the crash scene where Captain Kim Alan Granholm of the Thomson Township/Esko Volunteer Fire Department lost his life. Only the vehicles immediately behind the fire apparatus on the shoulder were on the scene at the time of the fatal crash........................22

Patrick Schneider, Charlotte Observer

Firefighters console one another at the funeral for Firefighter Joshua Earley................................54

ACKNOWLEDGEMENTS

TABLE OF CONTENTS

TABLE OF CONTENTS

TABLE OF CONTENTS

BACKGROUND

For 26 years, the United States Fire Administration (USFA) has tracked the number of firefighter fatalities and conducted an annual analysis. Through the collection of information on the causes of firefighter deaths, the USFA is able to focus on specific problems and direct efforts toward finding solutions to reduce the number of firefighter fatalities in the future. This information is also used to measure the effectiveness of current programs directed toward firefighter health and safety.

One of the USFA's main program goals is a 25-percent reduction in firefighter fatalities in 5 years and a 50-percent reduction within 10 years. The emphasis placed on these goals by the USFA is underscored by the fact that these goals represent one of the five major objectives that guide the actions of the USFA.

In addition to the analysis, the USFA provides a list of firefighter fatalities to the National Fallen Firefighters Foundation. If Memorial criteria are met, the fallen firefighter's next of kin, as well as members of the individual fire department, are invited to the annual Fallen Firefighters Memorial Service. The service is normally held at the National Emergency Training Center (NETC) in Emmitsburg, Maryland, during Fire Prevention Week. Additional information regarding the Memorial Service can be found on the Internet at http://www.firehero.org/ or by calling the National Fallen Firefighters Foundation at (301) 447-1365.

Other resources and information regarding firefighter fatalities, including current fatality notices, the National Fallen Firefighters Memorial database, and links to the Public Safety Officer Benefit (PSOB) program can be found at http://www.usfa.fema.gov/inside-usfa/ffmem.cfm.

INTRODUCTION

This report continues a series of annual studies by the USFA of onduty firefighter fatalities in the United States.

The specific objective of this study is to identify all onduty firefighter fatalities that occurred in the United States in 2002 and to analyze the circumstances surrounding each occurrence. The study is intended to help identify approaches that could reduce the number of firefighter deaths in future years.

In addition to the 2002 overall findings, this study includes a study of firefighters killed while responding in their personal vehicles and low cost steps that can be taken to prevent the loss of firefighter lives.

Who is a Firefighter?

For the purpose of this study, the term firefighter covers all members of organized fire departments in all States, the District of Columbia, and the Territories of Puerto Rico, the Virgin Islands, American Samoa, the Commonwealth of the Northern Mariana Islands, and Guam. It includes career and volunteer firefighters; full-time public safety officers acting as firefighters; State, Territory, and Federal government fire service personnel, including wildland firefighters; and privately employed firefighters, including employees of contract fire departments and trained members of industrial fire brigades, whether full- or part-time. It also includes contract personnel working as firefighters or assigned to work in direct support of fire service organizations.

Under this definition, the study includes not only local and municipal firefighters but also seasonal and full-time employees of the United States Forest Service, the Bureau of Land Management, the Bureau of Indian Affairs, the Bureau of Fish and Wildlife, the National Park Service, and State wildland agencies. The definition also includes prison inmates serving on firefighting crews; firefighters employed by other governmental agencies, such as the United States Department of Energy; military personnel performing assigned fire suppression activities; and civilian firefighters working at military installations.

What Constitutes an Onduty Fatality?

Onduty fatalities include any injury or illness sustained while on duty that proves fatal. The term "on duty" refers to being involved in operations at the scene of an emergency, whether it is a fire or nonfire incident; responding to or returning from an incident; performing other officially assigned duties such as training, maintenance, public education, inspection, investigations, court testimony, and fundraising; and being on-call, under orders, or on standby duty except at the individual's home or place of business. An individual who experiences a heart attack or other fatal injury at home as he or she prepares to respond to an emergency is considered on duty when the response begins. A firefighter that becomes ill while performing fire department duties and suffers a heart attack shortly after arriving home or at another location may be considered on duty since the inception of the heart attack occurred while the firefighter was on duty.

A fatality may be caused directly by an accidental or intentional injury in either emergency or nonemergency circumstances, or it may be attributed to an occupationally-related fatal illness. A common example of a fatal illness incurred on duty is a heart attack. Fatalities attributed to occupational illnesses would also include a communicable disease contracted while on duty that proved fatal when the disease could be attributed to a documented occupational exposure.

Injuries and illnesses are included even when death is considerably delayed after the original incident. When the incident and the death occur in different years, the analysis counts the fatality as having occurred in the year in which the incident took place.

One firefighter died in 2002 as the result of injuries he suffered in 1982. The USFA was notified of the deaths of two firefighters in 2001 that were not known or included in the firefighter fatality report for that year. One of the firefighters who died in 2001 was injured at a fire in 1997. For statistical purposes, each firefighter death is counted in the year in which the incident occurred. Information about these three deaths is included in the appendix of this report, but they are not addressed in the body of the report unless the death impacts retrospective statistical comparisons.

There is no established mechanism for identifying fatalities that result from illnesses such as cancer that develop over long periods of time, which may be related to occupational exposure to hazardous materials or products of combustion. It has proved to be very difficult over the years to provide a complete evaluation of an occupational illness as a causal factor in firefighter deaths due to the following limitations: the exposure of firefighters to toxic hazards is not sufficiently tracked, the often delayed long-term effects of such toxic hazard exposures, and the exposures firefighters may receive while off duty.

Sources of Initial Notification

As an integral part of its ongoing program to collect and analyze fire data, USFA solicits information on firefighter fatalities directly from the fire service and from a wide range of other sources. These sources include the Public Safety Officers' Benefit (PSOB) program administered by the Department of Justice, the National Institute for Occupational Safety and Health (NIOSH), the Occupational Safety and Health Administration (OSHA), the United States military, the National Interagency Fire Center, and other Federal agencies.

The USFA receives notification of some deaths directly from fire departments, as well as from such fire service organizations as the International Association of Fire Chiefs (IAFC), the International Association of Fire Fighters (IAFF), NFPA International, the National Volunteer Fire Council (NVFC), State fire marshals, State training organizations, other State and local organizations, fire service Internet sites, news services, and fire service publications. The USFA also keeps track of fatal fire incidents as part of its Major Fires Investigation Program and performs an ongoing analysis of data from the National Fire Incident Reporting System (NFIRS).

Procedure for Including a Fatality in the Study

In most cases, after notification of a fatal incident, initial telephone contact is made with local authorities by the USFA to verify the incident, its location, jurisdiction, and the fire department or agency involved. Further information about the deceased firefighter and the incident may be obtained from the chief of the fire department or his or her designee over the phone or by other data collection forms.

Information that is requested routinely includes NFIRS-1 (incident) and NFIRS-3 (fire service casualty) reports, the fire department's own incident reports and internal investigation reports, copies of death certificates or autopsy results, special investigative reports, police reports, photographs and diagrams, and newspaper or media accounts of the incident. Information on the incident may also be gathered from NFPA International, the USFA, or NIOSH reports on an incident.

After obtaining this information, a determination is made as to whether the death qualifies as an on duty firefighter fatality according to the previously described criteria. The same criteria were used for this study as in previous annual studies. Additional information may be requested, either by follow-up with the fire department directly, from State vital records offices, or other agencies. The determination as to whether a fatality qualifies as an on duty death for inclusion in this statistical analysis is made by the USFA. The final determination as to whether a fatality qualifies as a line-of-duty death for inclusion in the Fallen Firefighters Memorial Service is made by the National Fallen Firefighters Foundation.

2002 FINDINGS

One hundred firefighters died while on duty in 2002. While the number of deaths in 2002 is dramatically lower than the horrendous loss of firefighter lives in 2001, it is still an unacceptable level of loss. The level of firefighter fatalities trended downward in the early 1990s but has settled at somewhat higher levels in the latter part of the 1990s and into this century.

Figure 1. Onduty Firefighter Fatalities (1977-2002)

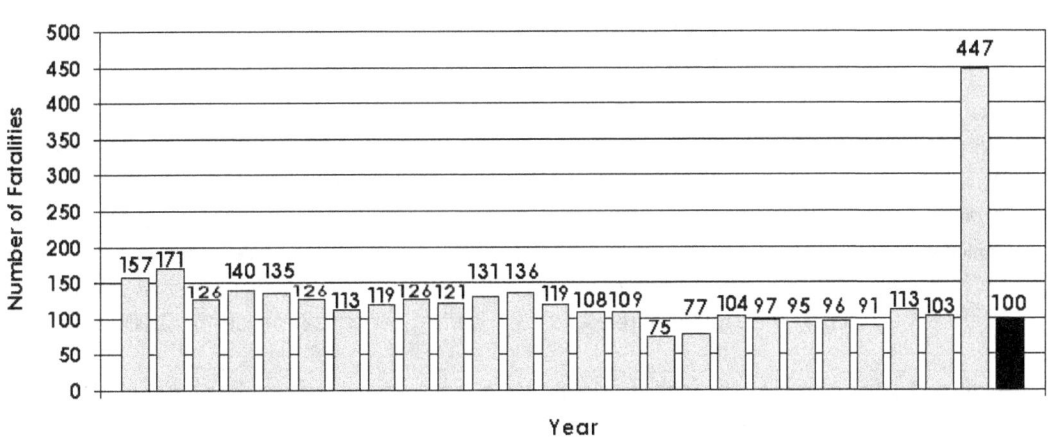

With the deaths of 100 firefighters in 2002, this is the fifth time in the last 10 years and the tenth time within the last 15 years when the total number of firefighter fatalities has reached or exceeded 100. The lowest years on record are 1992 with 75 fatalities and 1993 with 77 fatalities (Figure 1).

The 100 deaths resulted from a total of 84 incidents. There were nine multiple firefighter fatality incidents.

In 2001, 344 firefighters were killed as a result of the attacks on the World Trade Center (WTC) in New York City on September 11. When conducting multi-year comparisons of firefighter fatalities in this report, it may be necessary to set these deaths apart for illustrative purposes. This action is by no means a minimization of the supreme sacrifice made by these firefighters.

Seven Firefighters were murdered in 2002:

Six in arson-caused or suspicious fires

One at the hands of a gunman who had set a structure fire.

The median age for firefighters who died while on duty in 2002 was 41 years and 9 months.

Figure 2. Firefighter Fatalities per 100,000 Fire Incidents

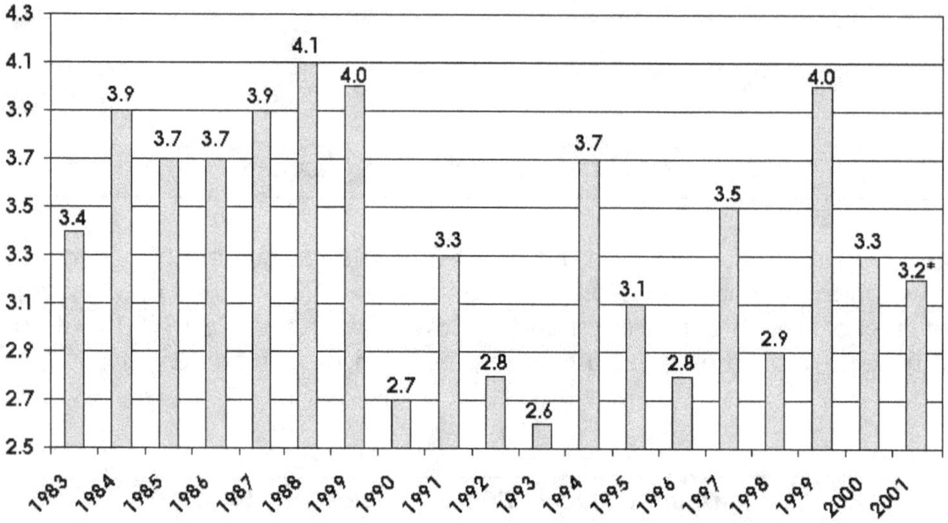

Bar chart showing values by year:
1983: 3.4, 1984: 3.9, 1985: 3.7, 1986: 3.7, 1987: 3.9, 1988: 4.1, 1989: 4.0, 1990: 2.7, 1991: 3.3, 1992: 2.8, 1993: 2.6, 1994: 3.7, 1995: 3.1, 1996: 2.8, 1997: 3.5, 1998: 2.9, 1999: 4.0, 2000: 3.3, 2001: 3.2*

*Does not include WTC

While the total number of firefighter fatalities has been decreasing over the past 20 years, the number of firefighter deaths per fire incident has actually risen. The chart above (Figure 2) compares the total number of firefighter fatalities each year that are associated with responses to fires and the total number of fire incidents reported by NFPA International through 2001 (2002 data is not yet available). Despite a downward dip in the early 1990s, the level of firefighter fatalities is back up to the same levels experienced in the 1980s. If the firefighter deaths at the WTC are included in the 2001 data, the number rises to 23.1 firefighter fatalities per 100,000 fires.

A retrospective study of firefighter fatalities reporting on firefighter fatalities over a 10-year period entitled "Firefighter Fatality Retrospective Study 1990-2000" is available at no cost from the USFA. The report may also be found on the internet at www.usfa.fema.gov/inside-usfa/fa-220.cfm.

Career and Volunteer Fatalities

Firefighter fatalities in 2002 includes 66 volunteer firefighters and 34 career firefighters (Table 1). Among the volunteer firefighter fatalities, 50 were from local or municipal volunteer fire departments, and 16 were seasonal or contract members of wildland fire agencies. All of the career firefighters that died were members of local or municipal fire departments. Ninety-five of the fatalities were men and 5 were women.

Table 1. Career vs. Volunteer Fatalities

TOTAL (443)	
Career (34)	**Volunteer (66)**
Metro Depts (18)	Suburban/Urban VFD (22)
Other Depts (16)	Rural VFD (28)
	Wildland Seasonal/Part-time (16)

Multiple Firefighter Fatality Incidents

The 100 fatalities resulted from 84 incidents. There were 9 multiple firefighter fatality incidents resulting in the deaths of 25 firefighters (Table 2).

In 2002, five firefighters died when the van in which they were riding was involved in a crash as it headed to join the firefighting effort in Colorado; three firefighters died in the crash of a wildland firefighting aircraft in California; three firefighters died in a structural collapse as they searched a building for victims in New Jersey; three California firefighters died when their wildland engine apparatus left the roadway and rolled down a steep embankment; three Oregon firefighters died as they fought a fire in a commercial building; two New York firefighters died as they advanced a hoseline into a fire-involved structure -- the floor collapsed and they fell into the fire area; two firefighters died during a structural fire-fight in Missouri; two wildland firefighters died in the crash of their airtanker in Colorado; and two Florida firefighters died during structural firefighting training.

Table 2. Multiple Fatality Incidents

Year	Number of Incidents	Total Number of Fatalities
2002	9	25
2001	8	362
2001 w/o WTC	7	18
2000	5	10
1999	6	22
1998	10	22
1997	8	17
1996	3	8

Wildland Firefighting Fatalities

The number of deaths associated with brush, grass, or wildland firefighting in 2002 was 23 (Table 3). In 2002, there were six firefighter deaths associated with wildland aircraft fire-fighting duties (two multiple fatality incidents claimed fivefirefighters and one crash claimed a single firefighter). This total includes fixed-wing aircraft and helicopters (Table 4).

> In 2002, there were no firefighter fatalities as the result of firefighters having their positions overrun by wildland fires.

Five firefighters based in Oregon died in the crash of a passenger van in Colorado; three California firefighters died when their engine left the roadway and rolled; two firefighters died while completing pack tests in California and Montana; two firefighters suffered heart attacks as they fought wildland fires; two firefighters died in apparatus crashes while responding to wildland fires; two firefighters were thrown from wildland firefighting vehicles and struck; and one firefighter died when he was struck by a falling tree.

Table 3. Fatalities Associated with Wildland Firefighting

Year	Total Number of Fatalities
2002	23
2001	15
2000	19
1999	28
1998	13
1997	10
1996	5

Table 4. Wildland Firefighting Aircraft Fatalities

Year	Total Number of Fatalities
2002	6
2001	6
2000	6
1999	0
1998	3
1997	5
1996	0

TYPE OF DUTY

In 2002, 73 onduty firefighter deaths were associated with emergency incidents (Figure 3). This includes all firefighters who died while responding to an emergency, while at the emergency scene, or while returning from the emergency incident. Nonemergency activities accounted for 27 fatalities. Nonemergency duties include training, administrative activities, or performing other functions that are not related to an emergency incident. A 7-year historical perspective concerning the percentage of firefighter deaths that occurred during emergency duty is presented in (Table 5).

Figure 3. Firefighter Fatalities by Type of Duty (2002)

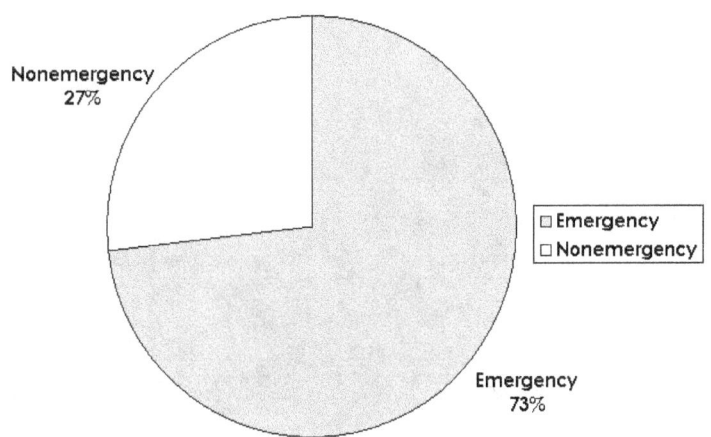

The number of fatalities by type of duty being performed in 2002 is shown in Table 6 and presented graphically in Figure 4. As in previous years, the largest number of deaths occurred during fireground operations. There were 45 fireground deaths, almost half of the total.

Table 5. Emergency Duty Firefighter Fatalities

Year	Percentage of All Fatalities
2002	73
2001	65
2001 w/WTC	92
2000	71
1999	87
1998	77
1997	81
1996	72

Table 6. 2002 Fatalities by Type of Duty

Type of Duty	Number of Fatalities
Fireground Operations	45
Responding/Returning from Alarm	13
Other Onduty Fatalities	14
Training	11
Nonfire Emergencies	12
After an Incident	5
Total	**100**

Figure 4. Fatalities by Type of Duty (2002)

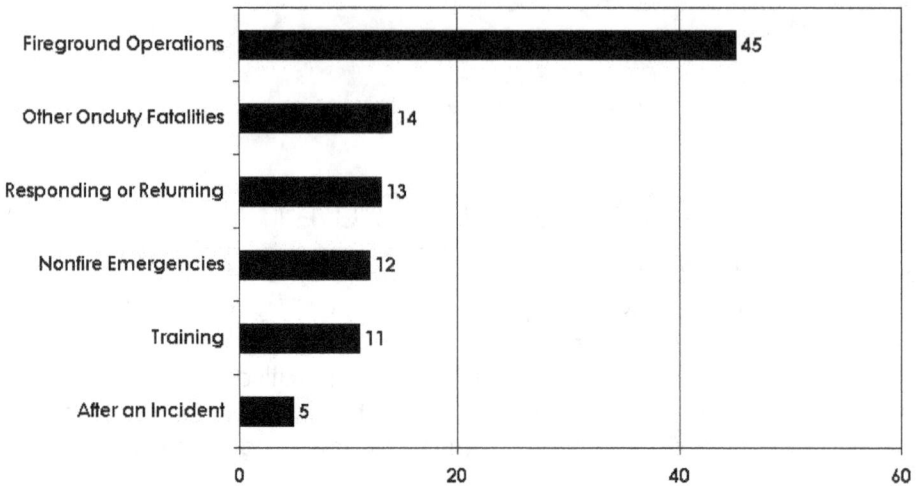

Fireground Operations — 45
Other Onduty Fatalities — 14
Responding or Returning — 13
Nonfire Emergencies — 12
Training — 11
After an Incident — 5

Fireground Operations

Many of the firefighting deaths in 2002 were related in some way to structural collapse. Three New Jersey firefighters were killed when the building that they were searching suffered a sudden catastrophic collapse; three Oregon firefighters died when a fire-weakened roof collapsed and trapped firefighters inside of the structure; two New York firefighters were killed as they entered a structure to fight the fire and were propelled into the fire area when the floor beneath them collapsed -- a similar incident took the life of a North Carolina firefighter; one firefighter in Indiana and one firefighter in Texas were killed after being buried in the debris of falling structures; a Pennsylvania firefighter was pinned by an initial collapse (firefighters were unable to move him) and then killed in a major collapse of a home being used for storage.

Firefighter Joshua Early was killed when the first floor of this structure collapsed and he fell into the fire-involved basement.

TYPE OF DUTY

Four firefighters were killed when they were trapped in the interior of burning structures. Two Missouri firefighters died of smoke inhalation in a burning commercial building, and one firefighter in Pennsylvania and one firefighter in Tennessee died of smoke inhalation in residential fires.

Heart attacks claimed the lives of 12 firefighters in 2002 while they were engaged in fire-related incidents. Ten of the heart attacks occurred at structure fires and 2 occurred at wildland fires.

A total of six firefighters were killed in firefighting aircraft crashes. Three firefighters were killed when the wings of their airtanker separated from the airframe and the plane crashed; two firefighters were killed when a wing of their airtanker broke off and the aircraft crashed; and one firefighter was killed in the crash of a helicopter.

Vehicle crashes at the fire scene took the lives of five firefighters in 2002. Three California firefighters were killed when their engine slid off the roadway and rolled into a stand of trees. One firefighter was killed when he was thrown from a brush truck after a collision -- he was severely burned after landing in the wildland fire flame front; and one firefighter was killed when he was thrown from the front bumper of a wildland apparatus when the apparatus was struck by another vehicle -- the firefighter's head was crushed by the forward movement of the apparatus after the crash.

One firefighter in New Mexico was killed on the scene of an incendiary structure fire when he was shot by the person who had started the fire; a firefighter in Minnesota was killed when he was struck by a vehicle while fighting a car fire on a highway; one firefighter in Michigan was killed when he was struck by someone jumping from a fire-involved apartment building, he suffered a head injury; a Massachusetts firefighter died of respiratory failure after fighting a structure fire; a firefighter in Colorado was killed by a falling tree; and an Iowa firefighter fell through a ventilation hole into the fire area.

Other Onduty Fatalities

Fourteen deaths occurred in 2002 during other onduty activities. Five wildland firefighters were killed when the van in which they were passengers was involved in a single-vehicle crash in Colorado; one firefighter in Connecticut, one firefighter in Pennsylvania, and, one firefighter in Tennessee died after becoming ill while on-duty; two firefighters died in their sleep while on duty, both due to heart problems; one firefighter suffered a Cerebral Vascular Accident (CVA/stroke) while exercising; one firefighter died in a crash when his vehicle was struck at an intersection as he drove to a meeting; one firefighter died when the tanker she was driving back to the station after repairs left the roadway and crashed; and one firefighter was killed while lighting fireworks for a community Independence Day celebration.

Responding/Returning

Thirteen firefighters died while responding to or returning from emergency incidents in 2002 (Table 7). Six firefighters died of heart attacks while in the fire station preparing to respond or while responding in their personal or fire department vehicles.

One firefighter was killed while responding to a brush fire in a pumper; she was a back seat passenger and was crushed when the apparatus was involved in a single-vehicle crash. The firefighter was wearing a seat belt.

One firefighter was killed when the tanker he was driving went off the right side of the roadway; he steered the apparatus back onto the road, but the apparatus began to skid and crashed off the right side of the road. This is an often-repeated fatal scenario for tanker fire apparatuses.

A firefighter that was killed in a vehicle collision while responding in 2002 had a blood alcohol level of .11, and was therefore legally intoxicated.

A new USFA publication, "Safe Operations of Fire Tankers" is now available. Further information about this publication may be found on the USFA Web site at www.usfa.fema.gov/inside-usfa/vehicle.cfm.

This photo depicts the aftermath of a tanker crash that killed Firefighter Jason Kevin Jackson on September 5, 2002.

Five firefighters died in 2002 while responding to emergencies in their personal vehicles. One firefighter failed to negotiate a turn and crashed; one firefighter's vehicle was involved in a crash when another vehicle crossed the center line of the roadway and struck the vehicle; one firefighter was killed when he was struck by a car while responding to an incident on his bicycle; two firefighters were killed while responding to the fire station in their personal vehicles. The subject of firefighter fatalities that occur while operating personally owned vehicles (POV's) will be discussed in detail in the special topics section of this report.

TYPE OF DUTY

Table 7. Fatalities While Responding to or Returning from an Incident

Year	Number of Fatalities
2002	13
2001	23
2000	19
1999	26
1998	14
1997	21
1996	22

Nonfire Emergencies

Twelve firefighters died while working at the scene of nonfire emergencies. This total includes three firefighters who suffered heart attacks as they provided for the safety of scenes in their roles as fire police officers; three firefighters who suffered heart attacks at Emergency Medical Services (EMS) scenes; one firefighter who became ill at an EMS scene and later died of a CVA; and one firefighter who suffered a heart attack during debris removal after a damaging storm passed through the area.

Three firefighters were struck at the scene of motor vehicle crashes. A Texas firefighter was killed by passing traffic when he arrived first on the scene of a crash and began treatment of the injured; a Florida firefighter was struck in the median of an interstate highway when a tractor-trailer truck failed to stop for the traffic backup associated with the original crash and entered the median area; and a Kansas firefighter was killed when he was struck by a fire engine as it arrived on the scene of a crash -- the engine had experienced a mechanical failure and was unable to stop.

A South Dakota firefighter was killed when he entered a molasses tank to rescue an incapacitated worker and also became incapacitated by the oxygen deficient and hydrogen sulfide contaminated atmosphere in the tank.

Training

Eleven firefighters died in 2002 while engaged in training activities (Table 8). Two Florida firefighters were killed in a structural firefighting training exercise; a Maryland firefighter died of hyperthermia after completing physical training during recruit school in extremely high heat; an Indiana firefighter drowned during rescue diver training; an Alabama firefighter was killed when a fire truck slipped into gear and struck him; a Kentucky firefighter was killed when the tanker he was driving back from a training exercise was struck by a train; and a New York firefighter was struck by a vehicle operated by an impaired driver as the firefighter loaded hose at the completion of a training exercise -- the driver had ignored traffic control devices.

Two firefighters died in 2002 while engaged in annual recertification tests as wildland firefighters, one from a heart attack in California and one from a CVA in Montana. One firefighter experienced a heart attack during classroom training in Virginia, and a North Carolina firefighter died of a heart attack during structural firefighting training.

Table 8. Fatalities During Training

Year	Number of Fatalities
2002	11
2001	14
2000	13
1999	3
1998	12
1997	5
1996	6

After an Incident

In 2002, 5 firefighters died after the conclusion of an emergency incident. All of the deaths were attributed to heart attacks. Three firefighters became ill as they completed paperwork at the conclusion of an EMS incident; one firefighter died in his sleep just after returning from an emergency incident; and one firefighter became ill at the scene of a structure fire, reported to the Incident Commander, went home, and was later discovered dead by friends.

Career, Volunteer, and Wildland Fatalities by Type of Duty

Figure 5 depicts career, volunteer, and wildland firefighter deaths by type of duty. Wildland career, wildland seasonal, and wildland contractor deaths were grouped together. As in past years, there were a disproportionate number of fatalities experienced by volunteer firefighters responding to and returning from alarms as compared to career and wildland firefighters. In 2002, a quarter of all volunteer firefighter deaths occurred while responding to or returning from

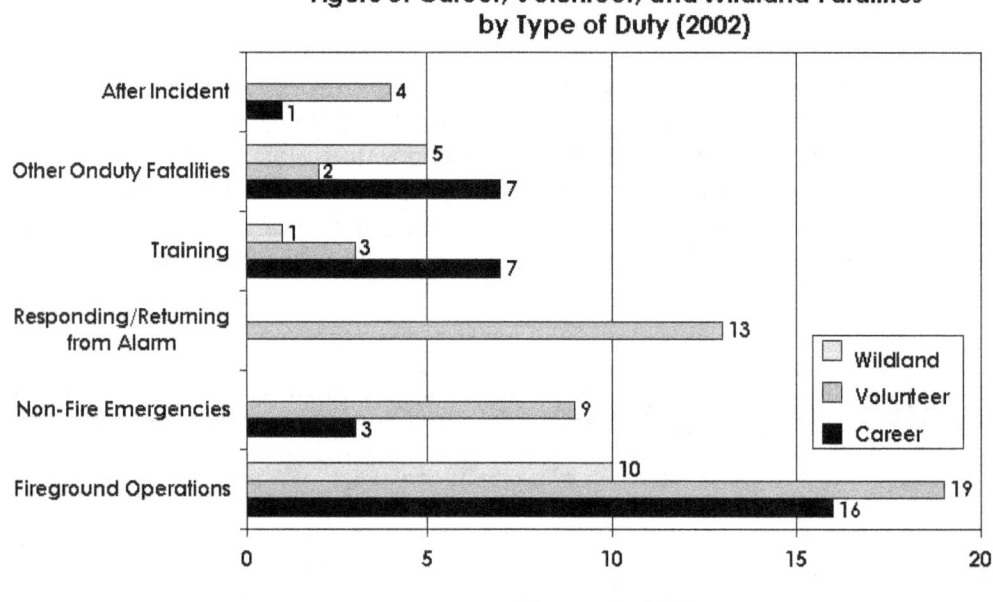

Figure 5. Career, Volunteer, and Wildland Fatalities by Type of Duty (2002)

Number of Fatalities

emergencies. In comparison, no career firefighter deaths and none of the wildland deaths occurred while responding or returning.

The large number of career firefighter deaths while on duty (but not involved in an incident or training activity) may be attributed to the fact that career firefighters are on duty for longer periods of time than volunteer firefighters. The onduty periods for volunteer firefighters generally are related to an emergency incident or other official functions such as training or fundraising. Some volunteer fire departments staff stations overnight (similar to a career department) but their numbers are small when compared to the total number of volunteer fire departments.

Type of Emergency Duty

In 2002, 73 firefighters died while engaged directly in the delivery of emergency services. This number includes deaths that were the result of injuries sustained on the incident scene or enroute to the incident scene, and firefighters that became ill on an incident scene and later died. It does not include firefighters who became ill or died while returning from an incident (such as a vehicle collision while returning from an incident). Figure 6 shows the number of firefighters killed in firefighting, emergency medical services, and other emergency incidents in 2002.

Fifty-eight firefighters were killed in relation to fires; 13 firefighters were killed in relation to EMS calls; and 2 firefighters were killed while engaged at a water main break in the community and assisting in the recovery after a tornado.

Figure 6. Type of Emergency Duty (2002)

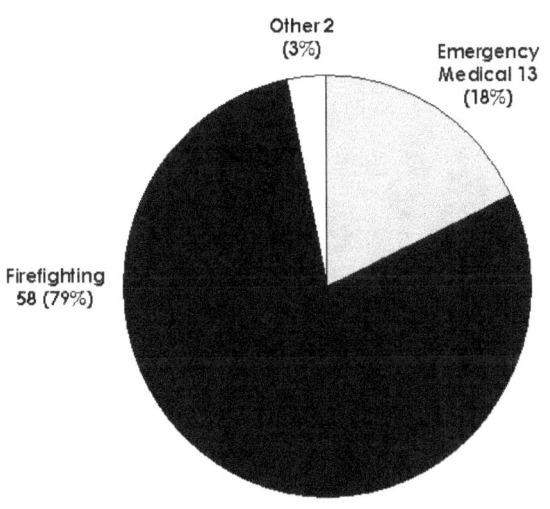

CAUSE OF FATAL INJURY

The term "cause of injury" refers to the action, lack of action, or circumstances that resulted directly in the fatal injury. The term "nature of injury" refers to the medical cause of the fatal injury or illness, often referred to as the physiological cause of death. A fatal injury usually is the result of a chain of events; the first of which is recorded as the cause.

In 2002, firefighters were killed when they ran out of air or were placed into fire areas by structural collapse. The cause of the fatal injury will be listed as "collapse" and the nature of the injury will be listed as "asphyxiation" or "burns."

Similarly, if a wildland firefighter was overrun by a fire and died of burns, the cause of death will be listed as "caught/trapped" by fire progress, and the nature of death will be "burns." This follows the convention used in the NFIRS casualty reports.

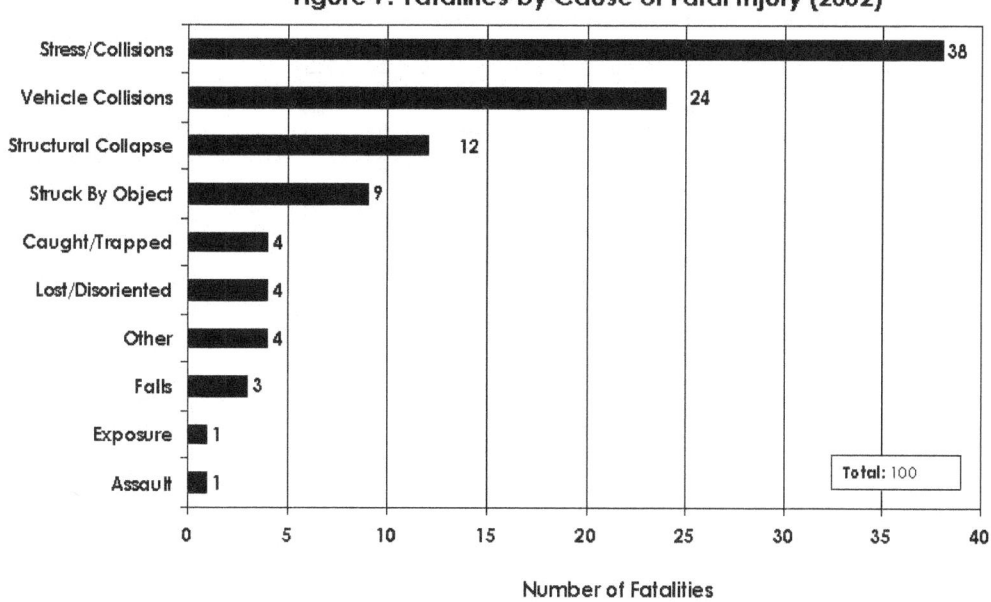

Figure 7. Fatalities by Cause of Fatal Injury (2002)

Stress or Overexertion

The largest cause of firefighter deaths is stress or overexertion, which was listed as the primary factor in 38 of the firefighter deaths in 2002 (Table 9).

Firefighting is extremely strenuous physical work and is likely one of the most physically demanding activities that the human body performs.

Most firefighter deaths attributed to stress result from heart attacks. Of the 38 stress-related fatalities in 2002, 34 firefighters died of heart attacks and 4 died of CVA's (stroke). Ten of the 38 deaths for which the cause of the fatal injury is listed as stress/overexertion occurred during nonemergency activities.

At 38 percent this is the lowest percentage of firefighter fatalities due to stress or overexertion, and the number in any year since at least 1996.

Top of your 2003/04 Reading List: *The South Beach Diet*, Author: arthur Agatston, M.D. "I am convinced that preventing most heart attacks and strokes is feasible today... the majority of heart attacks and strokes can be prevented."

Table 9. Fatalities Caused by Stress or Overexertion

Year	Number	Percent of Fatalities
2002	38	38%
2001	42	41.2*
2000	45	44%
1999	54	49%
1998	42	46%
1997	40	43%
1996	47	50.0%

(* Does not include the firefighter deaths of September 11, 2001, in New York City.)

Vehicle Collisions

The second leading cause of fatal injury for fire-fighters who died in 2002 was vehicle collisions. This cause is usually the second most common cause of firefighter fatalities.

Five wildland firefighters were killed when the van in which they were riding was involved in a single-vehicle crash; four firefighters were killed

In 2002, a firefighter was killed in the crash of a fire pumper responding to a brush fire. The speed of the apparatus was calculated at 74 mph in a 45 mph zone just prior to the crash.

in crashes involving their personal vehicles; three firefighters were killed when their engine apparatus slipped off a road and rolled; three firefighters were killed in crashes involving tankers, one each while responding, returning from training, and returning from maintenance; one firefighter was killed when he was struck while responding on his bicycle; one firefighter was killed when his departmental vehicle was struck as he traveled to a meeting; and one firefighter was killed when the fire apparatus in which she was a passenger was involved in a crash.

Six wildland aircraft firefighters were killed in 2002 in three separate incidents. Three firefighters were killed when the wings of their aircraft separated from the fuselage and the aircraft crashed; two firefighters were killed when one wing separated from the fuselage and the aircraft crashed; and a helicopter pilot was killed while fighting a fire in Colorado.

Firefighters console one another in front of the wreckage of the tanker which was operated by Firefighter Cassandra "Sandy" Myers Billings Powell at the time of her death. The apparatus crashed onto its roof after leaving the roadway.

Structural Collapse

An unusually large number of firefighters died due to structural collapse in 2002 (Table 10).

Three multiple firefighter fatality incidents took a total of eight firefighters, three in New Jersey, three in Oregon, and two in New York. An Indiana firefighter and a Texas firefighter were killed when they were buried by structural collapses. A firefighter in North Carolina fell into a fire-involved basement after entering a structure fire, and a Pennsylvania firefighter was first pinned in a structural collapse and then killed by a subsequent massive collapse.

Twelve percent of the 2002 firefighter fatalities were related to structural collapse. This is more than twice the level of any year since 1997.

Struck by Object

Being struck by an object was the fourth leading cause of fatal firefighter injuries in 2002. There were nine deaths in this category, including four firefighters who were struck by vehicles at emergency scenes. A North Dakota firefighter was killed when he was struck by an errant fireworks shell at a community Independence Day celebration where the fire department pro-

Table 10. Fatalities Caused by Structural Collapse

	Number	Percent of Fatalities
Year	12	12.07%
2002	4	3.9%*
2001	4	3.8%
2000	6	5.3%
1999	5	5.5%
1998	6	6.2%
1997	6	6.3%

(*Does not include WTC)

vided the fireworks; a military firefighter in Alabama was killed when he was struck by a fire apparatus that slipped into gear and rolled forward; a New York firefighter was killed as he helped to load hose after a training exercise and was struck by a vehicle that ignored traffic control devices; a fire victim jumped and struck a firefighter at a Michigan structure fire; and a falling tree killed a wildland firefighter who was engaged in removing dangerous trees from the forest line near a road.

The photo depicts the crash scene where Captain Kim Alan Granholm of the Thomson Township/Esko Volunteer Fire Department lost his life. Only the vehicles immediately behind the fire apparatus on the shoulder were on the scene at the time of the fatal crash.

CAUSE OF FATAL INJURY

Caught or Trapped

In 2002, four firefighters were killed when they were caught or trapped. Two Florida firefighters were killed when they were overcome by fire progress during a structural firefighting training session; an Indiana firefighter was killed when he was trapped underwater during dive rescue training; and a South Dakota firefighter died when he was trapped in a molasses tank while attempting to rescue a worker.

Lost or Disoriented

Four firefighters died in 2002 when they became lost or disoriented inside a structure fire and ran out of air. Two Missouri firefighters died in a fire in a commercial building; one firefighter in Pennsylvania and one firefighter in Tennessee died when they ran out of air in residential structure fires.

Other

Four firefighters died of causes that are not categorized above. Three died of natural cardiac diseases that were not stress related, and the cause of death for one firefighter could not be determined.

Falls

Three firefighters died in 2002 as the result of falls, the same number as in 2001. An Iowa firefighter died when he fell through a ventilation hole that had been cut into the roof of a burning structure; he fell into the fire area and died of asphyxiation. Firefighters in South Dakota and Texas were killed when they fell from exterior riding positions on fire apparatus during wildland firefighting duties. One firefighter died of burns and the other firefighter was crushed by the fire apparatus as it rolled forward.

Exposure

A Maryland firefighter died of hyperthermia when he became overheated during physical fitness training during the first few days of recruit school.

Assault

A New Mexico firefighter was killed by a gunman at the scene of a structure fire. The firefighter went to assist a burn victim who was located in a house near the burning structure. As the firefighter approached the house, he was shot and killed by the burn victim whom it was learned had set the original structure fire.

> From 1982 to 2002, 13 firefighters have been shot and killed by gunfire while on duty. There were two multiplefatality incidents. One of the 13 firefighters was struck when a rifle in a burning greenhouse discharged, the rest were at the hands of gunmen.
> (source: Hank Przybylowicz, Line of Duty Research Service)

NATURE OF FATAL INJURY

Figure 8 shows the distribution of the 100 deaths by the medical nature of the fatal injury or illness.

Figure 8. Fatalities by Nature of Fatal Injury (2002)

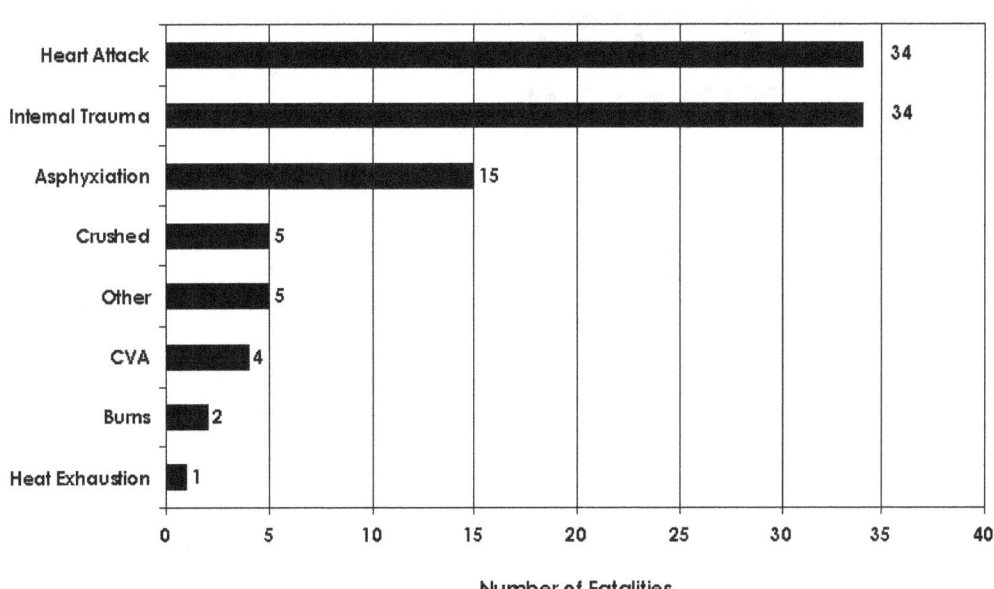

Heart Attack

In 2002, heart attacks and internal trauma were the nature of death for 34 firefighters. Heart attacks are usually the leading nature of firefighter deaths. Figure 9 provides a detailed break-down of heart attacks by type of duty.

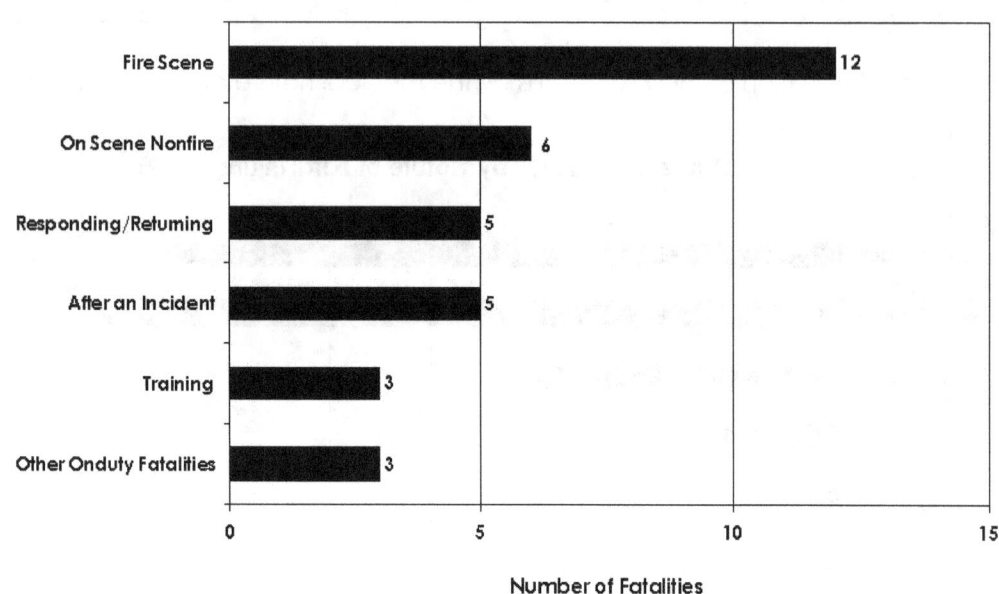

Figure 9. Heart Attacks by Type of Duty (2002)

Number of Fatalities

Twelve of the heart attacks occurred at the fire scene; 6 occurred at nonfire emergencies; 5 occurred as firefighters were responding to an emergency or returning from an emergency; 5 occurred after the conclusion of an incident; and 3 occurred during other duty-related activities. Three fatal heart attacks occurred during training, down from 10 in 2001 and 7 in 2000.

Internal Trauma

In 2002, internal trauma was tied as the nature of death that took the highest number of firefighter's lives, responsible for 34 deaths (Table 11). Twenty-three of these traumatic deaths occurred where firefighters were occupants of motor vehicles. Five firefighters died in Colorado when their van was involved in a single-vehicle collision; a total of six firefighters perished in aircraft crashes; four firefighters died in crashes that involved their personal vehicles; three firefighters died when their pumper slid off a California road and rolled; three firefighters were killed in separate tanker crashes; one firefighter was killed when the pumper in which she was riding was involved in a crash; and one firefighter was killed when his fire department vehicle was struck as he drove to an association meeting.

Four firefighters were struck by vehicles at the scene of motor vehicle crashes and one firefighter was killed when he was struck at the scene of a car fire.

A New Mexico firefighter was fatally shot during a structure fire; a Texas firefighter died of traumatic injuries resulting from a structural collapse; a Colorado firefighter was killed by a falling tree; a North Dakota firefighter was killed by an exploding fireworks shell; a Pennsylvania firefighter was killed in a collision as he was responding on his bicycle; and, an Alabama firefighter was killed when a fire truck slipped into gear and struck him.

NATURE OF FATAL INJURY

Table 11. Internal Trauma Fatalities

Year	Number of Fatalities
2002	34
2001	28*
2000	36
1999	25
1998	27
1997	32
1996	32

(*Does not include WTC)

Asphyxiation

Asphyxiation was the third leading medical reason for firefighter deaths in 2002, responsible for 15 deaths (Table 12). Twelve firefighters died of smoke inhalation after being involved in structural firefights, including training. A Pennsylvania firefighter died when he was unable to breathe due to the pressure of a structural collapse; a firefighter suffocated while trying to make a rescue from a molasses tank; and a firefighter in Indiana died during dive rescue training.

Table 12. Fatalities Due to Asphyxiation

Year	Number of Fatalities
2002	15
2001	18
2000	13
1999	16
1998	15
1997	15
1996	5

Crushed

In 2002, five firefighters died when they were crushed. Three New Jersey firefighters died when the residential structure that they were searching suffered a catastrophic collapse. An Indiana firefighter was crushed in the collapse of a commercial building, and a Texas firefighter was killed when he was crushed by the front wheel of a fire truck after he had fallen off the truck.

Other

Five firefighters were killed in situations where the nature of their fatal injuries do not fit into any of the categories. Three of the firefighters died of natural or hereditary heart disease; one firefighter died of respiratory failure after being exposed to smoke at a fire; and the cause of one firefighter's death could not be established.

CVA

Four firefighters died of CVA's (strokes) in 2002. One firefighter suffered a CVA while exercising at the fire station; one firefighter suffered a CVA as the result of being struck by someone who jumped from a structure fire in a residential occupancy; one firefighter suffered a CVA while performing a forestry pack test; and one firefighter suffered a CVA at the scene of a medical emergency.

Burns

Two firefighters died in 2002 where their deaths were primarily attributed to burns. Many times firefighters that die in structural fires suffer burns, but the burns are not the primary cause of death or the burns occur after the firefighter has died. A North Carolina firefighter fell into a burning basement and received severe burns. The firefighter lived for 2 days prior to dying from complications from his burns. A South Dakota firefighter was severely burned when he fell from the bed of a brush truck when the vehicle was struck. The firefighter fell into the flame front and received third-degree burns over 80 percent of his body. He lived for 5 days prior to his death from complications.

Heat Exhaustion

A Maryland firefighter was beginning his third day of recruit firefighter training. The day began with a series of physical fitness activities in extreme heat (code red) conditions. During a run returning to the training facility, the firefighter collapsed and died of hyperthermia.

FIREFIGHTER AGES

Figure 10 shows the percentage distribution of firefighter deaths by age and nature of the fatal injury. Table 13 provides counts of firefighter fatalities by age and the nature of the fatal injury.

Figure 10. Fatalities by Age and Nature (2002)

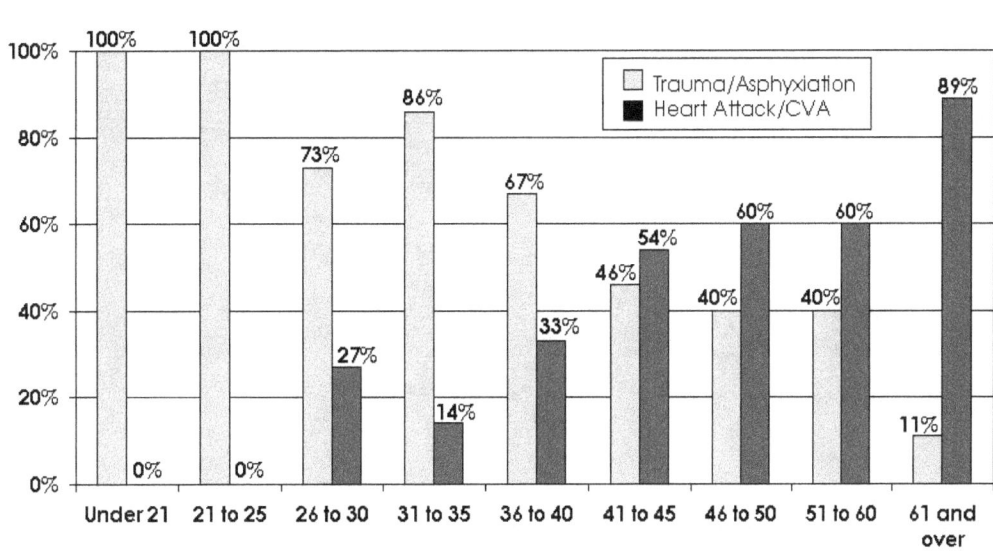

Firefighter Age

Table 13. Firefighters' Ages and Nature of Fatalities

	AGE								
	Under 21	21 to 25	26 to 30	31 to 35	36 to 40	41 to 45	46 to 50	51 to 60	61+
Non-Trauma	0	0	4	1	4	7	6	12	8
Trauma	10	4	11	6	8	6	4	8	1
Total	**10**	**4**	**15**	**7**	**12**	**13**	**10**	**20**	**9**

As in most years, younger firefighters were more likely to have died as a result of traumatic injuries such as injuries from an apparatus accient or after becoming caught or trapped during firefighting operations. Stress plays an increasing role in firefighter deaths as age increases.

The median age for firefighters who suffered fatal heart attacks or CVA's on duty in 2002 was 51 years and 3 months of age. Firefighters that died from heart attacks and CVA's ranged in age from 27 to 76.

The youngest firefighter killed in 2002 was Christopher Kangas of Pennsylvania at age 14, the oldest was Fire Police Captain Harold Coons of New York at 76.

FIXED PROPERTY USE FOR STRUCTURAL FIREFIGHTING FATALITIES

There were 27 firefighter fatalities in 2002 where the firefighters became ill while on the scene or while engaged in structural firefighting and the fixed property use is known. Table 14 shows the distribution of these deaths by fixed property use. As in most years, residential occupancies accounted for the highest number of these fireground fatalities, with 21 deaths.

Table 15 shows the number of firefighter deaths in residential occupancies for the last 6 years. Residential occupancies usually account for 70 to 80 percent of all structure fires and a similar percentage of the civilian fire deaths each year. Historically, the frequency of firefighter deaths in relation to the number of fires is much higher for nonresidential structures.

Table 14. Structural Firefighting Fatalities by Fixed Property Use

Fixed Property Use	Number	Percent
Residential	21	78%
Commercial	6	22%

Table 15. Firefighter Fatalities in Residential Occupancies

Year	Number of Firefighter Fatalities
2002	21
2001	17
2000	21
1999	23
1998	17
1997	16
1996	19

TYPE OF ACTIVITY

Figure 11 shows the types of fireground activities firefighters were engaged in at the time they sustained their fatal injuries or illnesses. This total includes all firefighting duties such as wildland firefighting and structural firefighting. In 2002 there were a total of 45 firefighter deaths on the fireground.

Figure 11. Fatalities by Type of Activity (2002)

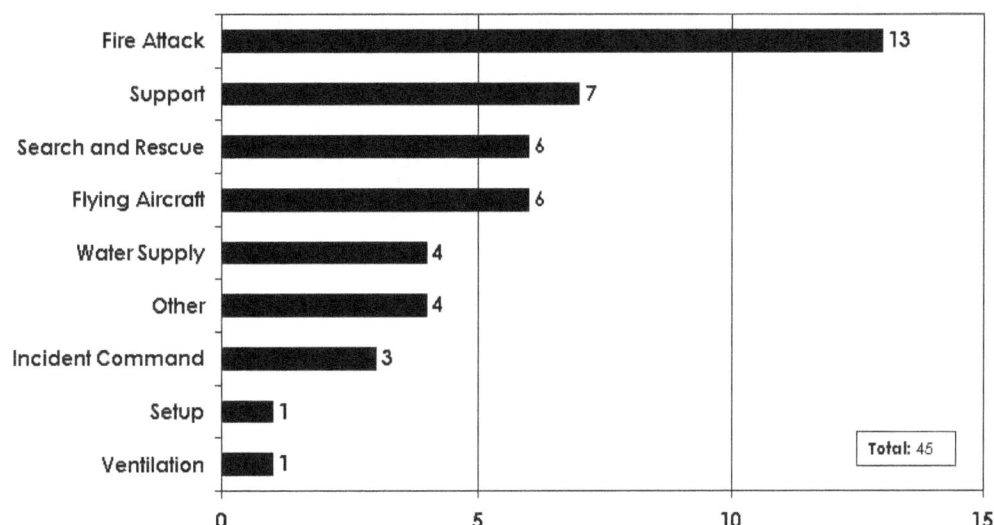

Fire Attack

Thirteen firefighters were killed as they engaged in direct fire attack, such as advancing or operating a hoseline at a fire scene. In years past, most fireground firefighter deaths occur while the firefighter is engaged in fire attack (see Table 16).

Three Oregon firefighters died while fighting a structural fire when a roof collapse occurred and trapped them; two New York firefighters died when a floor collapsed and they fell into a burning basement; a North Carolina firefighter also died when a floor collapsed and he fell into a burning basement; two firefighters suffered heart attacks as they fought fires; two firefighters fell from brush firefighting apparatus due to vehicle collisions and subsequently died; two firefighters died of smoke inhalation after they became lost or trapped inside burning structures; and one firefighter was pinned by a structural collapse and died while engaged in fire attack.

Table 16. Fatalities While Engaged in Fire Attacks

Year	Number of Fatalities
2002	13
2001	13
2000	13
1999	16
1998	18
1997	21
1996	9

Support

Seven firefighters were killed in 2002 as they supported firefighting efforts. A Texas firefighter was killed in a structural collapse as he assisted at the scene of a structure fire; a New Mexico firefighter was killed by a gunman at the scene of a structure fire; three firefighters died as they assisted with operations at structure fires; a Minnesota firefighter died when he was struck by a vehicle at the scene of a roadside car fire; and a wildland firefighter was killed when he was struck by a falling tree.

Search and Rescue

In 2002, six firefighters died while engaged in search and rescue operations at the scene of structure fires. Three New Jersey firefighters died when the building that they were searching collapsed; two Missouri firefighters were killed as they searched a commercial building for occupants and firefighters; and a Michigan firefighter died as the result of injuries he received while attempting to rescue trapped building occupants over ground ladders - a fire victim jumped from the building and struck the firefighter.

Flying Aircraft

Six firefighters died while flying firefighting aircraft. Three firefighters died in a California crash; two died in a Colorado crash; and the pilot of a helicopter was killed when his aircraft experienced a mechanical failure and crashed.

Water Supply

Four firefighters died in 2002 while engaged in water supply duties. All four deaths were heart-related. Three occurred at structure fires and one occurred at a wildland fire.

Other

Four firefighters were killed performing activities that are not classified. Three California wildland firefighters were killed when their pumper left the roadway and rolled. They were patrolling the perimeter of a fire to protect against fire spread. A Nebraska firefighter collapsed at the scene of a brush fire and died. No further information on this incident or the circumstances surrounding the death was available.

Incident Command

Two firefighters died in 2002 as the result of heart attacks suffered as they commanded structural fires. A Massachusetts firefighter died from complications from smoke inhalation that he suffered as he was in command of operations on the scene of a structure fire.

TYPE OF ACTIVITY

Setup

An Indiana firefighter was killed as he set up equipment to begin a defensive attack on a fire in a commercial building. He was struck as the building collapsed.

Ventilation

An Iowa firefighter was killed when he fell through the roof of a fire-involved residence. The firefighter had been engaged in roof ventilation duties and fell into the fire area as he and his partner attempted to leave the roof.

TIME OF INJURY

The distribution of all 2002 firefighter deaths according to the time of day when the fatal injury occurred is illustrated in Figure 12 (two incident times were not reported).

Figure 12. Fatalities by Type of Injury (2002)

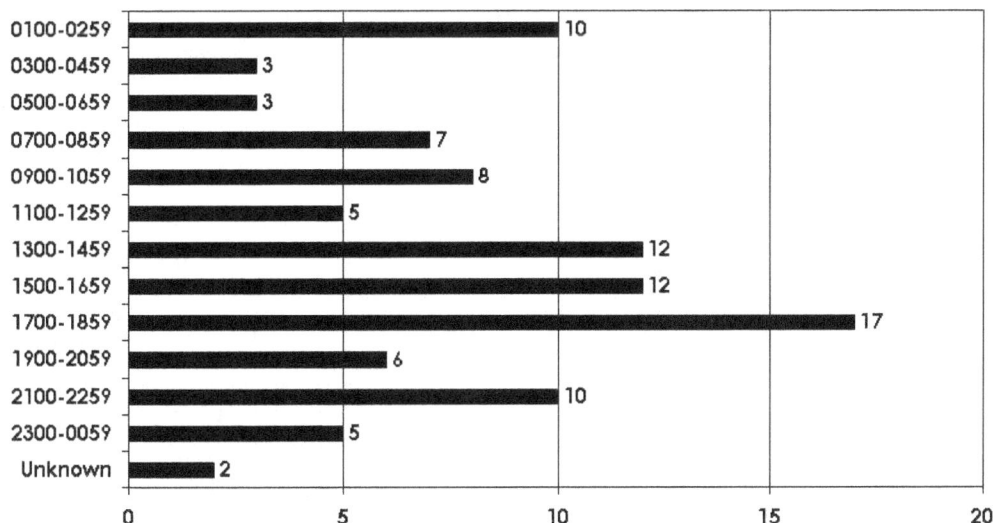

MONTH OF THE YEAR

Figure 13 illustrates firefighter fatalities by month of the year. Firefighter fatalities were highest in July due to a number of wildland firefighting deaths.

Figure 13. Fatalities by Month of the Year (2002)

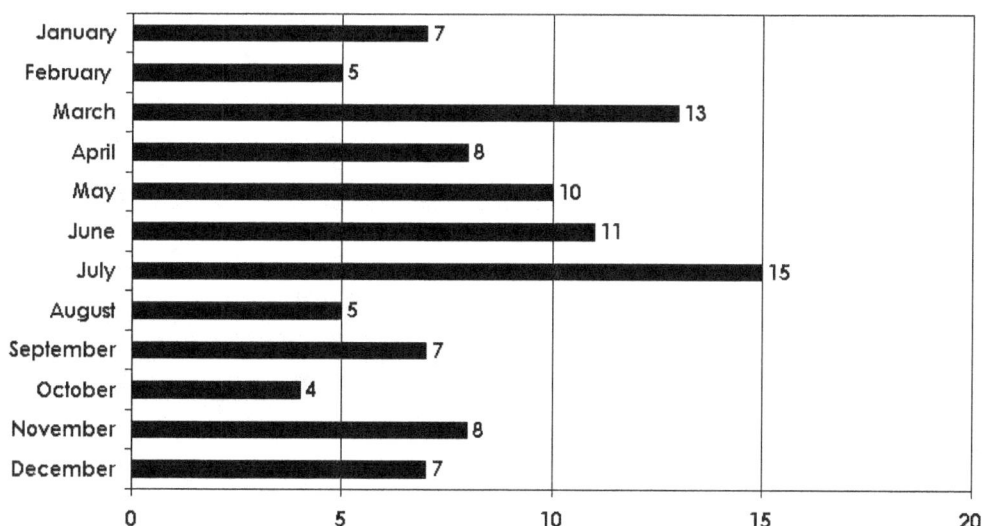

STATE AND REGION

The distribution of firefighter deaths by State is shown in Table 17. Firefighters based in 35 States died in 2002. Eight Oregon-based firefighters were killed. Five Oregon-based firefighters died in a van crash in Colorado, and three Oregon firefighters died fighting a structure fire in a commercial building.

The highest number of firefighter deaths occurred in Colorado. Nine firefighters died while on duty in Colorado in 2002. This is due to the extremely severe wildland fire season that took place in that State and two multiple firefighter fatality incidents that took a total of seven lives.

Figure 14 provides information on the ratio of firefighter fatalities per million population in each region.

Figure 14. Firefighter Fatalities by Region

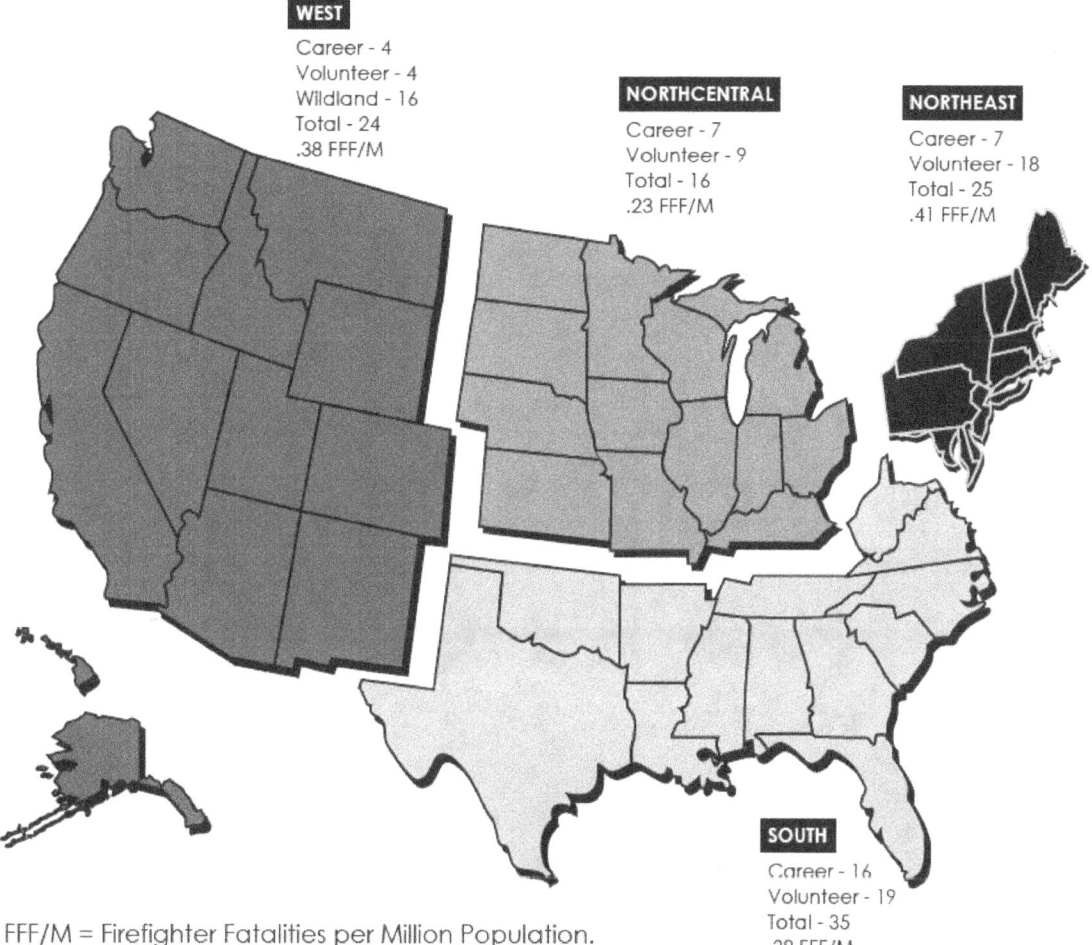

WEST
Career - 4
Volunteer - 4
Wildland - 16
Total - 24
.38 FFF/M

NORTHCENTRAL
Career - 7
Volunteer - 9
Total - 16
.23 FFF/M

NORTHEAST
Career - 7
Volunteer - 18
Total - 25
.41 FFF/M

SOUTH
Career - 16
Volunteer - 19
Total - 35
.39 FFF/M

FFF/M = Firefighter Fatalities per Million Population.

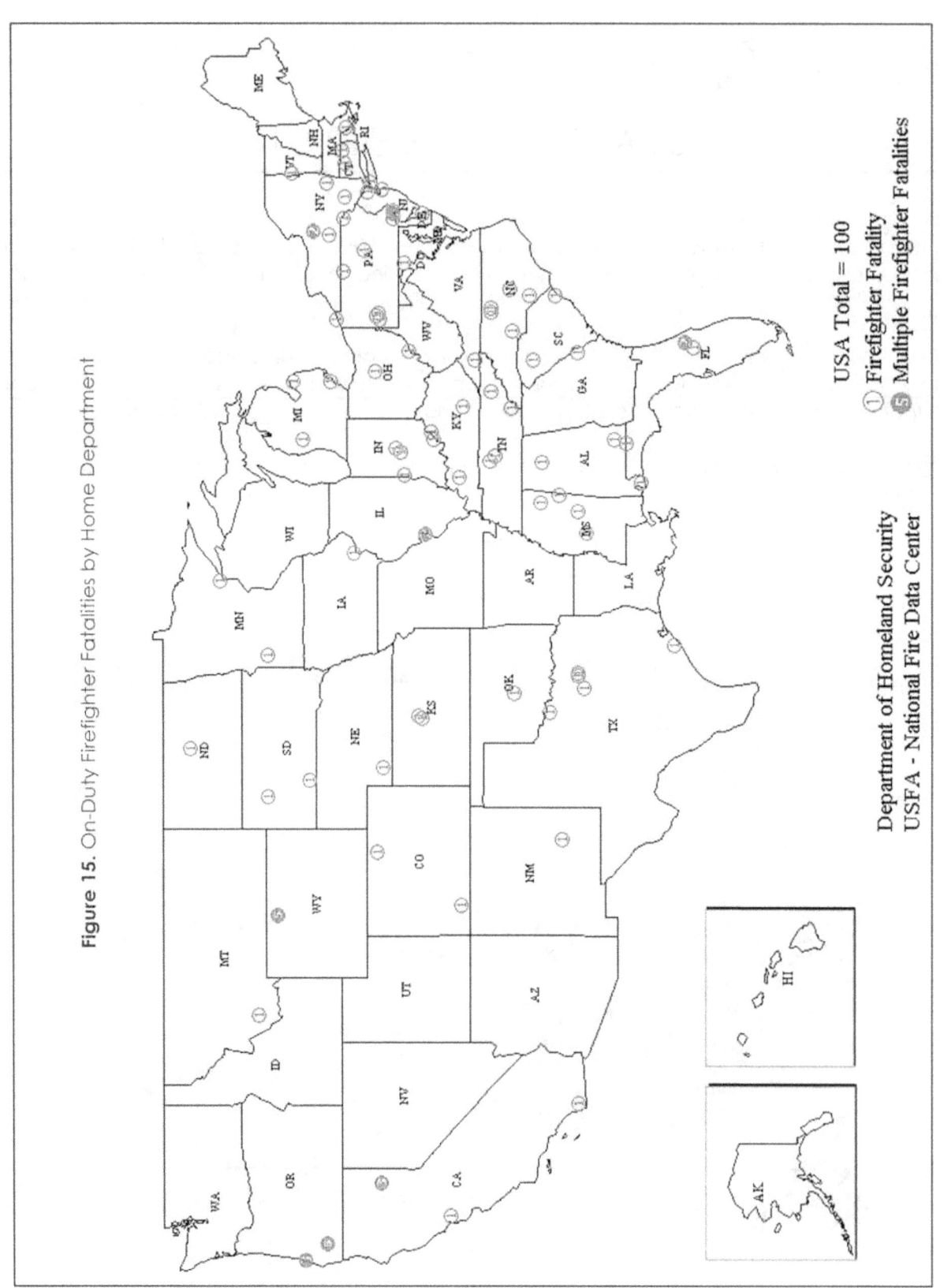

Figure 15. On-Duty Firefighter Fatalities by Home Department

USA Total = 100
① Firefighter Fatality
⑤ Multiple Firefighter Fatalities

Department of Homeland Security
USFA - National Fire Data Center

Figure 16. On-Duty Firefighter Fatalities by Incident Location

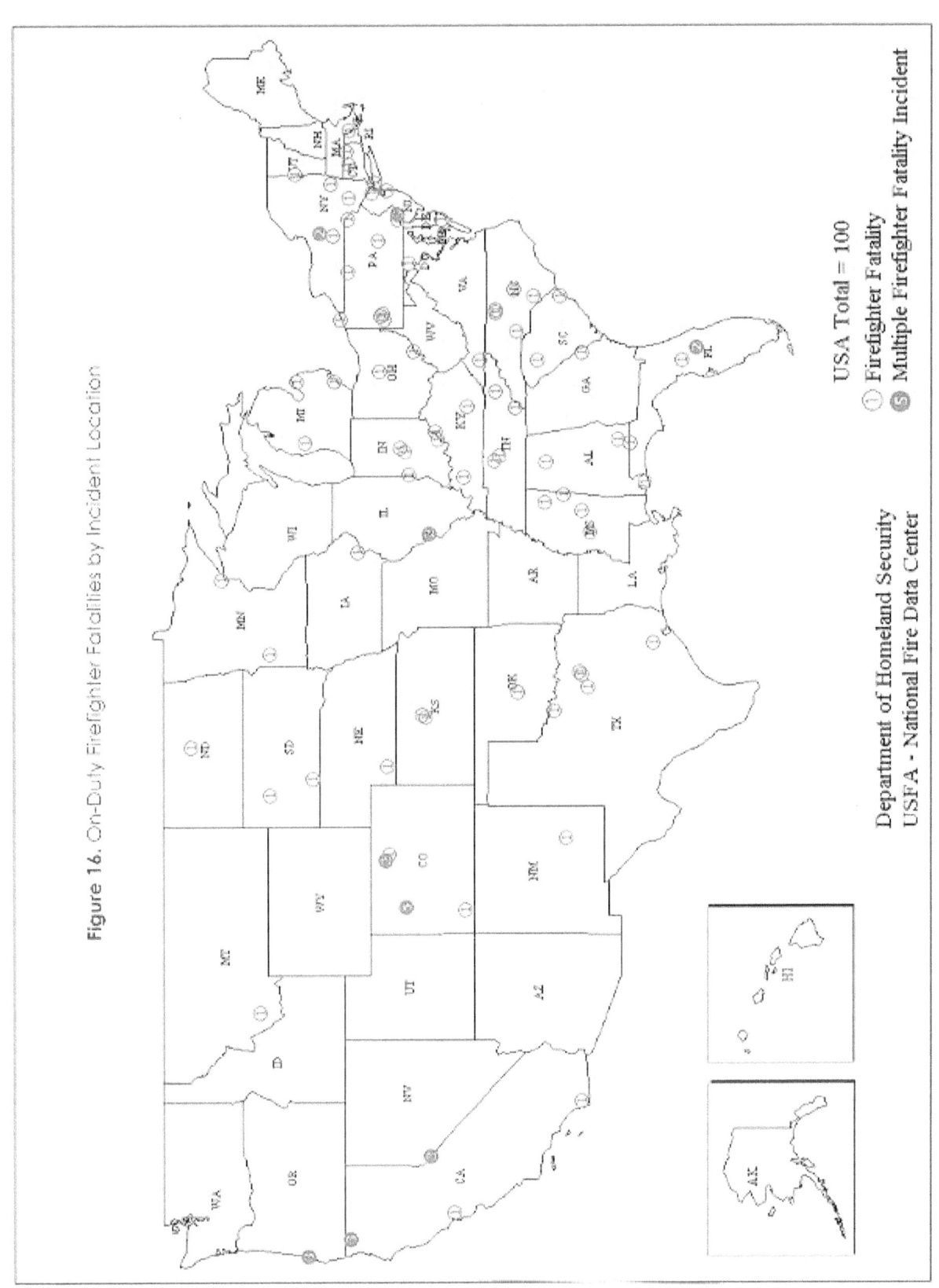

USA Total = 100
① Firefighter Fatality
⑤ Multiple Firefighter Fatality Incident

Department of Homeland Security
USFA - National Fire Data Center

Table 17. Fatalities by State

3	Alabama	3%
5	California	5%
2	Colorado	2%
3	Connecticut	3%
1	Delaware	1%
4	Florida	4%
1	Georgia	1%
1	Iowa	1%
4	Indiana	4%
2	Kansas	2%
2	Kentucky	2%
1	Massachusetts	1%
1	Maryland	1%
3	Michigan	3%
2	Minnesota	2%
2	Missouri	2%
4	Mississippi	4%
1	Montana	1%
5	North Carolina	5%
1	North Dakota	1%
1	Nebraska	1%
5	New Jersey	5%
1	New Mexico	1%
6	New York	6%
2	Ohio	2%
1	Oklahoma	1%
8	Oregon	8%
7	Pennsylvania	7%
2	South Carolina	2%
2	South Dakota	2%
5	Tennessee	5%
5	Texas	5%
1	Virginia	1%
1	Vermont	1%
5	Wyoming	5%

This list attributes the deaths according to the State in which the fire department or unit is based, as opposed to the State in which the death occurred. They are listed by those States for statistical purposes and for the National Fallen Firefighters Memorial at the National Emergency Training Center.

STATE AND REGION

ANALYSIS OF URBAN/RURAL/SUBURBAN PATTERNS IN FIREFIGHTER FATALITIES

The United States Bureau of the Census defines "urban" as a place having a population of at least 2,500 or lying within a designated urban area. Rural is defined as any community that is not urban. Suburban is not a census term but may be taken to refer to any place, urban or rural, that lies within a metropolitan area defined by the Census Bureau, but not within one of the central cities of that metropolitan area.

Fire department areas of responsibility do not always conform to the boundaries used for the census. For example, fire departments organized by counties or special fire protection districts may have both urban and rural coverage areas. In such cases, it may not be possible to characterize the entire coverage area of the fire department as rural or urban, and firefighter deaths were listed as urban or rural based on the particular community or location in which the fatality occurred.

The following patterns were found for 2002 firefighter fatalities (Table 18). These statistics are based on answers from the fire departments and, when no data from the departments were available, the data are based upon population and area served as reported by the fire departments.

Table 18. Fatalities by Coverage Area Type

Coverage Area	Fatalities
Urban/Suburban	53
Rural	33
Federal or State Parks/Wildland	14
Total	**100**

CONCLUSION

The year 2002 was another year of unacceptable loss for the fire service in the United States. The loss of a firefighter while on duty clearly is a direct loss for the family and coworkers of the deceased firefighter. The loss, however, is also felt within the community where the loss occurred and within the international fire service family.

In early 2003, when preliminary information regarding firefighter fatalities in 2002 was known, U.S. Fire Administrator R. David Paulison said: "The American Fire Service suffered another staggering year of loss in 2002. As the Nation's first responders, firefighters across the United States put their lives in danger every day. The United States Fire Administration is committed to helping improve firefighter safety to prevent these tragedies from occurring in the future."

The fire service has an unpaid debt to each of the firefighters that died in 2002 and before -- the prevention of future firefighter deaths.

This report contains two special topics. The first looks at an issue that mainly impacts volunteer firefighters -- personal vehicle crashes. Each year, firefighters die in single vehicle crashes as they respond to the fire station or to an incident scene in their personal vehicles. The key to the prevention of these deaths is training and the development and proper administration of procedures that require a slower approach and the use of seatbelts.

The second special topic area presents some inexpensive ways to make an immediate impact on the safety of firefighters. Topics include roadside safety, medical exam information, and the prevention of deaths caused by firefighters riding on exterior positions on wildland apparatus.

SPECIAL TOPICS

PERSONALLY OWNED VEHICLES

A Tragic Record

Firefighter deaths while operating Personally Owned Vehicles (POV's) is not a recent development. In 1993, 18 firefighters died while on duty responding to and returning from alarms. Tragically, 9 of the 18 firefighters died in their POV's, representing the leading type of vehicle that firefighters are operating when they are killed.

POV-response-related fatalities have remained at alarmingly high levels. This special topic report will examine case studies from 1997 through 2002 to highlight causative factors and to offer recommendations to prevent future deaths. Only knowledge of the problem and positive steps taken in advance of the response can stem this alarming tide of firefighter fatalities.

Twenty-five firefighters died in POV crashes from 1997 through 2002.

High speed was cited as a contributing factor in almost all of the POV crashes examined for this report. One firefighter was estimated to have been going almost twice the posted 40 mile-per-hour speed limit responding back to the fire station to retrieve his Personal Protective Equipment (PPE) after having missed the initial page for a mutual-aid response to a lumber yard fire.

Lack of seatbelt use. The lack of seatbelt use still presents a big problem. At least 6 of the 25 POV-related firefighter fatalities were listed as not wearing seatbelts at the time of the accident. It is likely that more than six firefighters were not using this important piece of safety equipment but police reports on some incidents are not clear on this issue.

Rollovers. There were six POV rollover fatalities. At least four POV operators were not wearing seatbelts at the time of their rollovers. What proved to be interesting was the type of vehicle and one of the vehicle's accessories. At least three of the rollover vehicles were Sport Utility Vehicles (SUV's) and at least three of the rollover vehicles were equipped with sunroofs. In one incident, a firefighter was responding in his personally owned Chevy Tahoe to the fire station. Responding down a downgrade, the Tahoe hit a patch of ice and began to slide. The Tahoe then left the roadway and rolled over three times. During one of the rollovers, the firefighter was ejected through the sunroof and suffered a fatal head injury. Excessive speed was also a contributing factor to this crash.

Intersections. In the past 5 years, eight firefighters died in POV intersection crashes. At least three of these crashes involved the use of colored lights, whether they were courtesy blue lights or red lights and a siren on the POV's.

Alcohol use. Several POV operators had high blood alcohol concentrations and would be considered legally intoxicated in most States. Firefighters who consume alcohol should not respond to emergencies.

Young members of the fire department merit special attention. Younger members may use bicycles or other nonstandard means of transportation and their driving skills are not as advanced as older members.

Head-on crashes. Head-on collisions account-ed for five POV firefighter fatalities during the period. Two firefighters died when their motorcycles were involved in collisions in the performance of their duties. One junior firefighter was killed when he ran a stop sign, and was struck by a vehicle as he was responding on his bicycle.

Family members as passengers. Another troubling trend emerged during the research for this report. Three firefighters killed while driving their POV's had family members, including children, responding with them at the time of their collisions. This practice is unsafe and should not be allowed by fire department policy.

The Need for an Emergency Response? Finally, most of the firefighters killed in their POV's were responding to calls that do not fit the national definition of a "True Emergency." According to the United States Department of Transportation Emergency Vehicle Operators Course instructor's manual: "a true emergency is a situation in which there is a high probability of death or serious injury to an individual or significant property loss, and action by an emergency vehicle operator may reduce the seriousness of the situation." Most calls for assistance are **not** true emergencies, but present fire department driving attitudes do not reflect this fact -- particularly when it involves POV operation.

Recommendations

- Firefighters should be trained in safe POV driving techniques before the first response. No first responder should be allowed to respond to any call without some form of effective driver training. This training should include instruction in State motor vehicle laws related to POV use in response to emergencies, driver safety, seatbelt use, safe parking, and vehicle placement.

- POV response to emergencies should be based upon a written procedure. The procedure should include guidance on when to respond, how to respond safely, and specific response requirements regarding speed, intersections, seatbelt use, inclement weather, and limited visibility situations (dark, fog, etc.).

- POV response policies must be enforced. If an officer or a member of the fire department witnesses another member responding in a manner that is not consistent with the procedure, it must be addressed through the chain of command. If violations of the policy are not corrected, the policy is a false sense of security. Some components of the policy deserve zero tolerance with noncompliance. These could include seatbelt use, running red lights, and excessive speed.

- Vehicle-specific training should be provided. Firefighters who operate specialized or unusual POV's should receive instruction on their operating characteristics. Instruction on the operating characteristics of SUV's should be provided to those who operate them.

• **Alcohol and drug impairment.** Firefighters need to consider themselves out-of- service if they have consumed alcohol or taken drugs.

• **Control the need for emergency response.** All fire department response policies need to be structured around the Federal definition of a "True Emergency." Too many firefighters responding in their POV'S were killed while responding to nonemergency calls.

The risk to passengers must be considered. Firefighters need to consider themselves out-of-service if they have any nonfire department personnel in their POV's.

Fire Service Emergency Vehicle Safety Initiative
The *Fire Service Emergency Vehicle Safety Initiative* is a partnership effort of the USFA and the U.S. Department of Transportation (DOT)/National Highway Traffic Safety Administration (NHTSA), and the DOT/Intelligent Transportation Systems Joint Program Office.

The long-term goal of this project is to reduce the number of firefighters killed and injured responding to and returning from emergencies that account for the second highest number of onduty fatalities as well as reducing firefighter deaths and injuries from being struck while performing emergency operations on the roadway.

Further information about the National Emergency Vehicle Safety Initative may be found on the USFA Web site at www.usfa.fema.gov/inside-usfa/vehicle.cfm.

INEXPENSIVE LIFE-SAVING STEPS

There are a number of steps that can be taken by fire departments to immediately and inexpensively reduce the chances that a firefighter will be killed while on duty. The preceding special topic on POV crashes pointed out some of them. This section will illustrate some inexpensive steps that can be taken to reduce the number of firefighters who are injured and killed while on duty each year.

ROADSIDE SAFETY

As our highways and roads become more congested, the risks to firefighters operating in close proximity to traffic increase. Protection of firefighters operating on these scenes is the first step that must be addressed upon the arrival of firefighters on the scene.

Roadside Safety Basics

- Consider any moving vehicle a threat. Drivers have varying levels of capability and concentration. Until they have passed you, don't ignore them.

- Create a safety zone for firefighters to work within. Use law enforcement vehicles and fire apparatus as a shield from moving traffic.

- Wear retroreflective materials. Firefighter turnout gear contains retroreflective materials that will help firefighters to be seen while on the emergency scene. Keep PPE clean so that the retroreflective material can do its job.

- Purchase and wear safety vests. Class III vests are inexpensive and should be worn by every firefighter on the incident scene.

- Don't blind oncoming drivers. The warning lights on fire apparatus may blind or confuse approaching drivers. Turn off lights such as wig-wag headlights that cause problems. Limit the number of warning lights that are in operation.

- Give drivers early warning. Use traffic cones, law enforcement officers upstream in traffic, and flaggers to notify drivers that an emergency exists and that they need to slow down.

- Minimize the number of vehicles on the scene. When the scope of the emergency has been assessed, send units away from the scene that are not needed. Stage responding units off of the roadway in a parking lot or on-ramp until they are needed.

- Firefighters exiting an emergency vehicle need to ensure that it is safe to do so. Firefighters must watch for oncoming traffic and other hazards -- look before getting out of the vehicle.

- NFPA 1500, *Standard on Fire Department Occupational Safety and Health Program*, states that "Flourescent and retro-reflective warning devices such as traffic cones with DOT-approved retro-relective collars and DOT-approved retro-reflective signs station "EMERGENCY SCENE" (with adjustable directional arrows) and illuminated warning devices such as flares and/or other appropriate warning devices shall be used to warn oncoming traffic of the emergency operations and the hazards to members operating at the incident." (Used with permission from NFPA, 1 Batterymarch Park, Quincy, MA 02269-9101.)

In 2002, 2 firefighters died when they arrived on the scene of roadside emergencies prior to the arrival of any emergency vehicles. These situations are extremely hazardous and require the on-scene firefighter to assume that he or she cannot be seen by drivers. The natural inclination to concentrate on the victims of the crash at the expense of personal situational awareness or a form of "tunnel vision" must be guarded against.

Free information on the safety of responders when operating at roadside emergencies is available at www.respondersafety.com.

The site includes a link to order a videotape that can be used in training and a link to a comprehensive Standard Operating Procedure (SOP) on working near moving traffic. The SOP can be customized to fit the needs of your department and can be changed to include the name of your fire department.

EXTERIOR RIDING POSITIONS

Riding on the back step or sideboards of fire apparatus has been prohibited since the very first edition of NFPA 1500. This prohibition includes the practice of firefighters riding in exposed positions on wildland or brush fire apparatus.

In 2002, 2 firefighters were killed when they were thrown from exterior riding positions on wildland or brush firefighting apparatus. In both cases, the fire apparatus was struck by another vehicle, and the firefighter was thrown off the apparatus as a result of the impact. In one case, the firefighter was thrown into the flame front and horribly burned. In the other case, the firefighter was thrown in front of the fire apparatus and was crushed as the apparatus continued to move forward.

A simple and free tactical change could prevent these types of deaths from occurring in the future.

NFPA 1500 offers some specific advice on the safe way to conduct this type of operation without the need for the firefighter to ride in an exposed position on the apparatus. The advice is contained in the annex of the standard and is therefore not a requirement of the standard, just additional information for the user of the standard.

NFPA 1500 suggests that the safe way to move along a wildland fire line is to position two firefighters to one side of the apparatus in full view (ahead) of the driver. Each firefighter should be equipped with a hoseline and the apparatus operated in noninvolved areas. As the firefighters walk and fully extinguish the fire, the apparatus follows. The driver looks out for the firefighters and the firefighters look out for the driver and the apparatus.

MEDICAL EXAMINATIONS FOR FIREFIGHTERS

NFPA 1582, *Standard on Medical Requirements for Fire Fighters* and Information for Fire Department Physicians, require annual medical evaluations for all firefighters. The evaluation includes a review of the firefighter's medical and occupational experience for the year and assessments of the firefighter's height and weight, blood pressure, and heart rate and rhythm. The annual medical evaluation can be performed by a qualified person other than the fire department physician as long as their results are reviewed by the fire department physician.

More rigorous physical examinations are required every 3 years for firefighters up to age 29, every 2 years for firefighters between the ages of 30 and 39, and every year for firefighters age 40 and over.

NFPA 1582 contains a number of resources for the fire department and the physician performing these evaluations and examinations. The standard lists physical conditions that should pose a concern and provides guidance to physicians on the physical rigors of firefighting.

A copy of the standard can be purchased for approximately $30.00, including shipping. Fire department chief or administrative officers should make sure that their fire department physician has a copy of the standard. The minimal cost involved in having a few copies of the standard available to the members of the fire department could be more than overcome with the savings realized from avoiding a single injury.

It is also a good idea to make copies of the standard available for loan to all fire department members. Loaner copies of the standard can be taken by individual members to review with their personal physicians or provided to the physicians who perform physicals associated with their full-time jobs.

Medical information is a very private subject. The confidentiality of this information is protected by a number of laws and standards. Firefighters should be able to get medical advice from someone who understands the risks and rigors of the work of a firefighter -- full-time career and volunteer firefighters included.

Firefighters console one another at the funeral for Firefighter Joshua Earley.

SPECIAL TOPICS

APPENDIX - SUMMARY OF 2002 INCIDENTS

January 10, 2002 - 0230hrs
Steven M. Olander, Firefighter
Age 39, Career
Detroit Fire Department, Michigan

An arsonist spread flammable liquid on the stairs of an apartment building from the fourth floor to the second floor and started a fire. The fire progressed quickly and trapped a number of building occupants in their apartments.

Firefighter Olander and his ladder company were dispatched to the scene. They proceeded to the rear of the structure and were ordered to effect the rescue of victims trapped by the fire and awaiting rescue at windows on the fourth floor. Due to limited staffing, firefighters were unable to raise a 50-foot ground ladder, and electrical wires in the alley prevented the use of the aerial ladder. Firefighter Olander and two other firefighters were in the process of raising a 35-foot ground ladder to a fourth-story window. Firefighter Olander was footing the ladder when a fire victim jumped from the fourth story and struck a van and Firefighter Olander simultaneously.

Firefighter Olander was knocked to the ground but was able to get back on his feet and continue with rescue efforts. After the incident was concluded, Firefighter Olander complained of severe head, neck, and back pain. Firefighter Olander visited a medical facility at the end of his shift. He was treated and released. Over the next several weeks, Firefighter Olander continued to suffer severe headaches. On January 26, 2002, he collapsed and was admitted to a hospital. He remained in a coma until his death on February 7, 2002.

The fire eventually went to three alarms. Two residents of the apartment building were killed, a child was stillborn due to injuries received by the mother, and 11 other residents were injured.

Firefighter Olander's death was caused by a ruptured berry aneurysm, a form of cerebral aneurysm.

For additional information regarding this incident, please refer to NIOSH Fire Fighter Fatality Investigation and Prevention Program report F2002-14 (www.cdc.gov/niosh/face200214.html).

January 11, 2002 - 2215hrs
Richard Anthony Majors, Firefighter
Age 45, Career
Nashville Metro Fire Department, Tennessee

Firefighter Majors was on duty in his regularly assigned fire station. Firefighter Majors became ill and was suffering from respiratory distress. He was transported to the hospital where he subsequently died of a heart attack. The cause of death was listed as hypertension and coronary artery disease.

January 16, 2002 - 2001hrs
Robert W. Feeney, Firefighter
Age 41, Volunteer
Phil Daly Hose Company of the Long Branch Fire Department, New Jersey

Firefighter Feeney was in the process of meeting with members of his department's fire explorer program. During the meeting, Firefighter Feeney and other members of his department were dispatched to a second-alarm structure fire. Firefighter Feeney drove an engine to the scene. He was in the process of connecting a 5-inch supply line to his pump when he suffered a heart attack.

Other firefighters on the scene began cardiopulmonary resuscitation (CPR) immediately, and Firefighter Feeney was transported to a local hospital. Despite efforts to revive him, Firefighter Feeney was pronounced dead at the hospital. The cause of death was listed as occlusive coronary arteriosclerosis.

Phil Daly Hose Company #2, Long Branch Fire Department Web site --
www.fire-ems.net/firedept/view/LongBranch7NJ

January 19, 2002 - 1522hrs
Elmer Bell, Firefighter
Age 44, Career
Choctaw Fire Department, Mississippi

Firefighter Bell was participating in physical training in preparation for attendance at the Mississippi Fire Academy in Jackson. While engaged in stair climbing, Firefighter Bell became ill. He was transported to the hospital but died on January 21, 2002. The cause of death was listed as a CVA.

January 21, 2002 - 1613hrs
Thomas Evans Andersen, Assistant Chief
Age 48, Career
Surfside Beach Fire Department, South Carolina

Chief Andersen and the members of his department responded to a report of fire in a 3-story condominium. First-arriving firefighters found smoke on the first floor and active fire on the second floor. An attack line was deployed to the second floor for fire extinguishment and another line was deployed to look for fire extension. The fire was knocked down with no extension into the attic.

Chief Andersen was in the second floor landing overseeing operations after helping to extend the second line. He was wearing full structural firefighting protective clothing and an self-contained breathing apparatus (SCBA). He suddenly collapsed. Firefighters brought him down one floor and removed his protective coat and SCBA. CPR was initiated immediately.

Chief Andersen had a history of heart disease. He had recently completed a physical examination that would have resulted in his removal from emergency duty. The report of the physical examination was on the fire chief's desk and was not reviewed by the fire chief until after Chief Andersen's death. The fire department was fined by the Occupational Safety and Health Administration for allowing Chief Andersen to respond to emergencies although he was not physically capable of doing so.

The cause of death was listed as arteriosclerotic cardiovascular disease.

January 21, 2002 - 1100hrs
Dustin Michael "Dusty" Schwendeman, Firefighter
Age 26, Volunteer
Dunham Township Volunteer Fire Department, Ohio

Firefighter Schwendeman was responding in his personal vehicle to the fire station after his department was dispatched to a mutual-aid structure fire. Road conditions were wet with patches of ice.

As Firefighter Schwendeman responded on a downgrade, his vehicle hit a patch of ice and began to slide. Firefighter Schwendeman's Tahoe left the roadway to the right, struck a mailbox, struck an embankment, rolled three times, and landed back on its wheels. Firefighter Schwendeman was ejected through the vehicle's sunroof during one of the rollovers and suffered a severe head injury.

Firefighters from a neighboring department were dispatched to the crash. Firefighter Schwendeman was treated at the scene and transported to a local hospital. He was later transported by helicopter to a regional care facility. He died the following morning.

The cause of death was listed as a laceration of the brain and blunt trauma injuries. Firefighter Schwendeman was not wearing a seat belt at the time of the crash. The police report on the incident cited excessive speed as a contributing factor in the crash.

For additional information regarding this incident, please refer to NIOSH Fire Fighter Fatality Investigation and Prevention Program report F2002-04 (www.cdc.gov/niosh/face200204.html).

January 31, 2002 - 0600hrs
Thomas Earl Brooks, Master Firefighter
Age 48, Career
Lumberton Fire Department, North Carolina

Master Firefighter Brooks worked a regular 24-hour shift. During the shift, he participated in standard tasks such as checking equipment, flowing fire hydrants, training, and responding to two fires and one emergency medical incident. Firefighters returned from their last emergency response at approximately 2300.

When firefighters awoke the next morning, they found Firefighter Brooks in a chair in the station's lounge area. Firefighter Brooks had died some time during the night of a heart attack.

February 3, 2002 - 1000hrs
Louis Alfred Rickards, President
Age 55, Volunteer
Lewes Volunteer Fire Department, Delaware

President Rickards was traveling in a fire department vehicle to attend an executive meeting of the Delmarva Volunteer Fireman's Association. As President Rickards passed through Snow Hill, Maryland, he was involved in a crash.

President Rickards' vehicle was struck broadside by a vehicle that failed to stop at a traffic signal. President Rickards' vehicle rolled over and he was ejected. President Rickards was wearing a seatbelt. President Rickards was pronounced dead at the scene.

February 10, 2002 - 0920hrs
Rex Walter, Firefighter
Age 69, Volunteer
Napanoch Fire Company, New York

Firefighter Walter arrived at his fire station in response to the report of a chimney fire. As he donned his protective clothing, he collapsed of an apparent heart attack.

February 11, 2002 - 1855hrs
Vincent Llyonell Davis, Firefighter
Age 42, Career
Dallas Fire-Rescue, Texas

Firefighter Davis was assigned to fill in for a vacancy at Engine 33 on the day of the fatal incident. A fire was reported in a vacant apartment building that was undergoing renovation. Engine 33 initially was dispatched to move up to another fire station but was brought to the scene on the third alarm.

Upon their arrival on the scene, Firefighter Davis and the members of his crew checked for fire extension on the second floor of the southwest wing. After confirming that the attic of the involved wing and the attic of the southwest wing were not connected, Firefighter Davis and his crew returned to the ground floor. The captain of Firefighter Davis' company conferred with other officers in the area and the decision was made to deploy an additional handline. Firefighter Davis and his crew began to walk through a breezeway to a nearby engine company apparatus.

The captain was in the lead, Firefighter Davis and his driver were one arms length behind the captain, and a firefighter was one arms length behind Firefighter Davis and the driver. As the captain neared the end of the breezeway, a collapse occurred; Firefighter Davis and the driver were buried. Engine 33 had been on the scene for less than 8 minutes when the collapse occurred.

The driver was freed from the rubble after some difficulty, but the whereabouts of Firefighter Davis could not be confirmed. A boot was found in the rubble, and the search for Firefighter Davis began.

Firefighter Davis was found in a sitting position with his face down on his legs. He was wearing his SCBA but the unit was not activated. His Personal Alert Safety System (PASS) devices were also not activated. Firefighter Davis had no pulse. Debris was cleared and medical treatment began. CPR was initiated immediately and onscene paramedic firefighters provided Advanced Life Support (ALS) care. Care continued en route to the hospital. Firefighter Davis was pronounced dead at the hospital.

The elapsed time from the collapse to the removal of Firefighter Davis from the rubble was approximately 28 minutes. The fire was caused by the careless use of a construction torch. The cause of death was listed as blunt force trauma and traumatic asphyxiation.

For additional information regarding this incident, please refer to NIOSH Fire Fighter Fatality Investigation and Prevention Program report F2002-07 (www.cdc.gov/niosh/face200207.html).

Dallas Fire-Rescue Web site -- www.dallasfiredept.com

February 13, 2002 - 1830hrs
Robert Samuel "Bobby" Nichols, Jr., Lieutenant
Age 39, Volunteer
Loretto Volunteer Fire Department, Alabama

Lieutenant Nichols and members of his department responded to the scene of a water main break. The break was creating a traffic hazard as it washed out a nearby roadway.

Lieutenant Nichols had been on the scene for only a few minutes when he was struck with a heart attack. Firefighters on the scene immediately began treatment and an ambulance was called. Unfortunately, Lieutenant Nichols could not be revived.

Loretto Volunteer Fire Department Web site -- www.fire-ems.net/firedept/view/LorettoAL

February 18, 2002 - 1642hrs
Raymond Alvin Ebel, Fire Chief
Age 62, Paid-on-Call
Newaygo Fire Department, Michigan

Chief Ebel and the members of his department responded to a mutual-aid residential structure fire.

After the fire was under control, Chief Ebel collapsed of an apparent heart attack. Firefighters and paramedics on the scene provided aid immediately, and Chief Ebel was rushed to the hospital. Despite their efforts, Chief Ebel was pronounced dead at the hospital.

Chief Ebel had been a member of his department since 1966 and had been the fire chief for 4 years. He had a history of chest pain and bypass surgery.

March 1, 2002 - 1500hrs
Thomas Shane Murray, Firefighter
Age 21, Volunteer
Jefferson City Fire Department, Tennessee

Firefighter Murray and members of his department were dispatched to a report of a structure fire in a single-family residence. A fire inspector had discovered the fire and was about to report it when the incident was dispatched.

An engine company arrived on the scene, a water supply was established, and two attack lines were advanced. Firefighter Murray, who was a city employee, arrived in his city vehicle at this time. The fire chief and Firefighter Murray joined two other firefighters in the interior and completed a primary search of the structure. Finding an all-clear, the fire chief and Firefighter Murray retrieved a hoseline from the front entrance of the house for fire control. A positive-pressure fan was placed at the front entrance of the structure and windows were broken out for ventilation. A backup line from another engine company was advanced into the interior.

The hoselines were not having much effect on the fire and the second hose-line became useless when the booster tank on the second engine ran out of water. A third line was deployed but interior conditions continued to worsen. Based on his view of the exterior of the structure, the Incident Commander (IC)ordered an evacuation. Due to problems with the IC's radio, firefighters inside the structure did not hear the order.

Conditions continued to worsen inside the structure, and the fire chief ordered everyone to exit the structure. A firefighter and a lieutenant were first out the door and made it to the front yard. The chief, however, had difficulty exiting and collapsed just after stepping outside the structure. He could not get up and was helped to safety by other firefighters. An accountability report was taken, and Firefighter Murray was found to be missing.

The fire had progressed to the point that further entry into the structure was impossible. A deck gun was used to darken down the fire. Firefighters were able to see Firefighter Murray about 5 feet inside of the front door of the structure. He was removed to the street where EMS treatment was initiated. CPR was started and continued as Firefighter Murray was transported to the hospital. He was pronounced dead upon arrival.

Firefighter Murray was wearing and using his SCBA and PASS device. The low air warning and PASS alert tone did not help in his discovery. The cause of death for Firefighter Murray was listed as asphyxiation with a carboxyhemoglobin level of 31.8 percent.

For additional information regarding this incident, please refer to NIOSH Fire Fighter Fatality Investigation and Prevention Program report F2002-12 (www.cdc.gov/niosh/face200212.html).

March 4, 2002 - 2152hrs
Richard James Dake, Firefighter
Age 48, Volunteer
LaGrange Fire & Rescue Department, Kentucky

Firefighter Dake was the driver and sole occupant of a 1,500-gallon tanker. Members of Firefighter Dake's department had just completed training and Firefighter Dake was assigned to drive the tanker back to the fire station.

Firefighter Dake was preparing to cross an unguarded railroad crossing and failed to hear the horns or see an approaching train. The train struck the tanker on the right side just forward of the passenger door. Firefighter Dake was ejected through the passenger window. He had not been wearing a seatbelt.

Firefighters who had witnessed the crash located Firefighter Dake and began treatment immediately. Firefighter Dake was airlifted to a trauma hospital where he was pronounced dead due to multiple blunt force trauma.

For additional information regarding this incident, please refer to NIOSH Fire Fighter Fatality Investigation and Prevention Program report F2002-10 (www.cdc.gov/niosh/face200210.html).

March 4, 2002 - 1222hrs
Joshua Brandon Earley, Firefighter
Age 23, Part-Time Paid
Harrisburg Volunteer Fire & Rescue Department, Inc., North Carolina

Firefighter Earley and the members of his engine company responded to the report of a structure fire. Their response was an automatic mutual-aid response into a neighboring fire district. When the first units arrived on the scene, they reported a working fire. Firefighters attempted to enter the structure but were driven back by intense heat.

Firefighter Earley was the first firefighter through the door at another entry point. He had advanced a 1-3/4-inch hoseline 4 to 5 feet inside of the structure when the floor collapsed. Firefighter Earley fell into the fire-involved basement. The Captain backing him up on the line was able to avoid the fall by grabbing the door frame and was assisted by other firefighters to the exterior.

Firefighter Earley was removed from the structure by other firefighters approximately 1 minute after he fell into the basement. Medical care was provided by firefighters and EMS workers on the scene. Firefighter Earley received second and third-degree burns over 87 percent of his body.

Firefighter Earley was airlifted from the scene and was later transferred to a regional burn treatment facility. He expired due to complications from his burns on March 6, 2002. Firefighter Earley was also a career firefighter for the Charlotte Fire Department.

The fire started due to ordinary combustibles being stored too close to a wood-burning stove.

For additional information regarding this incident, please refer to NIOSH Fire Fighter Fatality Investigation and Prevention Program report F2002-11 (www.cdc.gov/niosh/face200211.html).

Harrisburg Fire & Rescue Department Web site - www.fire-ems.net/firedept/view/Harrisburg3NC

March 7, 2002 - 2000hrs
John Evo "Gino" Ginocchetti, Firefighter/Paramedic
Age 41, Career
Manlius Fire Department, New York

Timothy John "TJ" Lynch, Firefighter/Paramedic
Age 28, Volunteer
Manlius Fire Department, New York

Firefighter/Paramedic Ginocchetti and Firefighter/Paramedic Lynch responded with 2 other firefighters in a ladder truck to a mutual-aid structure fire. Fire was reported in the basement of a house.

Upon their arrival at the scene, the Manlius truck company was ordered to the roof to ventilate. The hole produced heavy smoke and heat. After returning to the ground, the crew was directed to relieve a crew operating a handline in the garage area of the home. Firefighter/Paramedic Lynch took the nozzle and Firefighter/Paramedic Ginocchetti backed him up. The line was advanced from the garage into the mudroom of the house. As soon as the firefighters made entry into the structure, the floor beneath them failed and they fell into the fire area.

An officer entered the mudroom and encountered heavy smoke and heat. He was unaware that a collapse had occurred until he heard Firefighter/Paramedic Ginocchetti calling for help. The officer tried to grab hold and help Firefighter/Paramedic Ginocchetti back into the mudroom, but he was driven back by intense heat and fire. The officer received burns to his hands and face after his SCBA face piece was pulled off during the rescue attempt. Other firefighters also attempted to rescue Firefighter/Paramedic Ginocchetti but they too were driven back by fire progress.

The collapse made access to the firefighters impossible through any existing entrances. A hole was breached into the back basement wall and firefighters were able to remove debris and locate both firefighters. They were removed from the basement and transported to the hospital where they were pronounced dead. The cause of death for both firefighters was asphyxiation.

APPENDIX

Firefighter/Paramedic Ginocchetti had a carboxyhemoglobin level of 15 percent and the level in Firefighter/Paramedic Lynch's blood was not detected.

The total time that had passed from the collapse to the removal of both firefighters was approximately 3 hours.

The fire was caused by sparks from a grinder being operated by the homeowner.

Firefighter Lynch was a career member of the Village of Fayetteville Fire Department.

For additional information regarding this incident, please refer to NIOSH Fire Fighter Fatality Investigation and Prevention Program report F2002-06 (www.cdc.gov/niosh/face200206.html).

Manlius Fire Department Web site -- www.fire-ems.net/firedept/view/ManliusNY

March 16, 2002 - 0100hrs
Steven Louis Jones, Fire Chief
Age 46, Career
Roswell Fire Department, New Mexico

Chief Jones and members of his department were dispatched to a report of a structure fire. The first engine arrived on the scene and reported a fully involved structure. The IC learned that there was a burn victim in the house across the street from the burning structure, and a medic unit was directed to the house to initiate treatment. About the same time, Chief Jones arrived at the command post and asked how he could help. The IC asked Chief Jones to interview the burn victim prior to transport to the hospital.

Moments after Chief Jones left the command post, gunfire was heard on the fire scene. Firefighters were ordered by the IC to take cover and a personnel accountability report was ordered. All firefighters were accounted for except Chief Jones.

When Chief Jones had approached the house to talk with the burn victim, the man opened fire and struck Chief Jones and a paramedic employed by a private company that responded on the initial dispatch. Police officers were able to move Chief Jones and the paramedic to a shielded area. On scene efforts were diverted from the fire fight and treatment was provided to the injured.

Chief Jones suffered a severe gunshot wound to the head. He was treated on scene by his firefighters and local EMS workers. He was transported to the hospital and ended up at a regional hospital in Texas. He survived until March 26, 2002, when he died of complications of his injury.

The shooter, a man with a history of mental illness, had set the structure fire. He shot and killed Chief Jones, the paramedic, and a neighbor. He shot and injured the neighbor's 3-year-old boy, and took another 5-year-old child hostage prior to ending his own life. Chief Jones' wife is a firefighter with the Fort Worth Fire Department in Texas.

Roswell Fire Department Web site - www.roswell-usa.com/city/pubsaf.htm

March 16, 2002 - 0230hrs
Clarence Francis Birchmore, Fire Chief
Age 60, Volunteer
Whiting Volunteer Fire Department, Vermont

Chief Birchmore was responding in his personal vehicle to the report of a vehicle crash. About 100 yards from the scene, he suffered a heart attack. A Vermont State Trooper and an ambulance, who were also responding to the original incident, witnessed his vehicle pull off the road and stopped to help.

CPR was begun immediately and Chief Birchmore was transported to the hospital. He was pronounced dead after all efforts failed to bring him back.

March 16, 2002 - 1041hrs
Leo Leon Swank, Lieutenant
Age 50, Volunteer
Jefferson Township Bellville Fire Department, Ohio

Lieutenant Swank and members of his fire department responded to the report of a chimney fire in a residence. The fire was controlled and units assembled and headed back to the firehouse.

Lieutenant Swank was a passenger in the jump seat behind the cab on a pumper apparatus. As the pumper made a turn, Lieutenant Swank suffered a heart attack and fell from the vehicle. Paramedics riding in a vehicle behind the pumper provided immediate medical aid. Lieutenant Swank was transported to the hospital but he did not survive.

At the time of his death, Lieutenant Swank's brother and one of his sons were firefighters with his department and two other sons were members of the department's explorer program.

March 18, 2002 - 1215hrs
Joan Esther Spear, Senior Engine Boss
Age 45, Wildland Full-Time
Montana Department of Natural Resources & Conservation, Fire & Aviation Management, Dillon Unit

Senior Engine Boss Spear was preparing for an annual recertification pack test. The test requires the firefighter to carry a 45-pound pack a distance of 3 miles within 45 minutes.

Senior Engine Boss Spear embarked on a practice walk carrying a 25-pound pack. Well into the walk, a passerby saw her struggle and fall to the ground. The passerby called 9-1-1. Despite the efforts of local EMS providers, Senior Engine Boss Spear died.

The cause of death for Senior Engine Boss Spear was a CVA. She was 4 days short of her 46th birthday.

APPENDIX

March 20, 2002 - 1459hrs
Adam Lee Weisenberger, Private
Age 19, Volunteer
Gluckstadt Volunteer Fire Department, Mississippi

Firefighter Weisenberger and members of his department were dispatched to a two-car crash with injuries on a local interstate highway. At the time of the dispatch, the area was experiencing heavy rain and thunderstorms. Firefighter Weisenberger was the first firefighter on the scene when he arrived in his personal vehicle.

Firefighter Weisenberger observed that two cars were involved and put on medical examination gloves. As he was talking with one of the victims, a passing vehicle collided with the vehicle occupied by the victim. The victim's vehicle, pushed by the secondary collision, knocked Firefighter Weisenberger into the traffic lane where he was run over by a passing truck. Witnesses said that Firefighter Weisenberger tried to dodge the approaching car but was unsuccessful.

Firefighter Weisenberger was thrown over 75 feet and came to rest in the median. Arriving firefighters provided treatment to Firefighter Weisenberger, and he was transported to the hospital. He was later pronounced dead at the hospital. The cause of death was listed as craniocerebral trauma.

Firefighter Weisenberger's father is the emergency management director for Madison County, Mississippi.

March 20, 2002 - 1500hrs
Bernes J. "Bernie" Schutte, Firefighter
Age 69, Volunteer
Palisade Volunteer Fire Department, Nebraska

Firefighter Schutte and other members of his department were working on the scene of a 4,500-acre wildland fire. Firefighter Schutte collapsed in the presence of other firefighters and medical treatment was begun immediately.

Firefighter Schutte was transported to a local hospital where he was pronounced dead upon arrival. The cause of death was listed as an acute myocardial infarction (heart attack) due to coronary artery disease.

Firefighter Schutte had been a member of the Palisade Volunteer Fire Department for 42 years. He had a history of heart problems.

March 25, 2002 - 2105hrs
Allen Frye, Captain
Age 31, Volunteer
Roslyn Rescue Hook and Ladder Company #1, New York

Captain Frye and the members of his department had just completed a rescue drill at an abandoned restaurant. Captain Frye and another firefighter were at the rear of a pumper loading hose.

A car operated by a driver who was impaired by a combination of alcohol and anti-depressant drugs approached the pumper. The driver drove around cones and barrels marking the training site and struck the two firefighters at the rear of the pumper. Captain Frye was killed and the other firefighter was severely injured.

The Roslyn Rescue Hook and Ladder Company #1 lost two members on September 11, 2001, in the attacks on the World Trade Center. Brothers Tom and Peter Langone were killed in the attacks. One brother was a police officer and the other an FDNY firefighter.

The driver of the vehicle was later charged with second-degree vehicular manslaughter and driving while intoxicated.

Roslyn Rescue Hook and Ladder Company #1 Web site -- www.roslynrescue.org

March 27, 2002 - 2045hrs
Fred McNeil Johnson, Firefighter
Age 71, Volunteer
Abingdon Volunteer Fire Department, Virginia

Firefighter Johnson and other members of his department attended a training session at the fire station. After the class was completed, Firefighter Johnson began to experience chest pains as he walked downstairs.

Firefighter Johnson was transported to the hospital for treatment. The next morning he was in the process of being transferred to another hospital when he went into cardiac arrest. He was pronounced dead in the emergency room.

April 1, 2002 - 1850hrs
Jackie Eli Ellington, Jr., Firefighter
Age 19, Paid-on-Call
Newcastle Fire Department, Oklahoma

The Newcastle Fire Department is a combination fire department. Career and paid-on-call firefighters staff the fire station with the presence of the paid-on-call firefighters required from 1900 until 2200. Firefighter Ellington was not scheduled to work the night of April 1, 2002. Additional staffing was needed at the fire station, so Firefighter Ellington was paged to report for duty.

Firefighter Ellington was responding to the fire station in his personal vehicle when he was involved in a crash. As Firefighter Ellington's vehicle reached the crest of a hill, a vehicle traveling in the opposite direction crossed the center line and struck Firefighter Ellington's vehicle.

The Newcastle Fire Department was dispatched to the scene. The fire chief was first on the scene and found that the driver of the other vehicle was killed, a passenger in the other vehicle was injured, and that Firefighter Ellington was killed also.

Firefighter Ellington was wearing his seatbelt but was partially ejected from his vehicle. The cause of death was listed as multiple trauma.

April 6, 2002 - Time Unknown
Mark David Mansfield, Firefighter
Age 30, Career
Overland Park Fire Department, Kansas

Firefighter Mansfield was on duty in his fire station. He went to bed at approximately 2300hrs but failed to respond to a dispatch at 0546hrs. Firefighters found Firefighter Mansfield in his bed; he had no pulse and was not breathing.

Firefighters, including paramedics, began treatment immediately, and Firefighter Mansfield was transported to the hospital. He was pronounced dead upon arrival.

The cause of death was listed as a cardiac arrhythmia due to mitral valve prolapse.

April 6, 2002 - 2350hrs
Kevin Leo Baker, Firefighter
Age 39, Volunteer
Mid-North Volunteer Fire Department, Texas

Firefighter Baker and members of his department responded to the report of a shooting. The apparatus occupied by Firefighter Baker staged a safe distance from the incident scene pending the arrival of law enforcement.

After the scene was secured by law enforcement, fire and EMS vehicles moved to the scene and provided treatment to the shooting victim. Recent rains had saturated the ground and most areas were muddy. The mud was so deep that it forced responders to pull their feet up with each step.

Firefighter Baker helped carry the ambulance gurney into the involved residence and then helped carry the gurney and the shooting victim back to the ambulance. Firefighter Baker then rode in his engine company about a quarter of a mile to a medical helicopter landing zone. As the shooting victim was removed from the ambulance to be loaded into the helicopter, Firefighter Baker stepped to the side of the ambulance and collapsed.

Firefighter Baker was treated at the scene and during the 28-minute ride to the hospital by ground ambulance. After efforts in the emergency room failed to revive him, he was pronounced dead.

The cause of death was listed as hypertensive arteriosclerotic cardiovascular disease.

April 7, 2002 - 1647hrs
Edna Faye Bishop, Firefighter
Age 29, Volunteer
Bon Secour Fire/Rescue, Alabama

Firefighter Bishop was a rear-seat passenger in a 4-door commercial pumper responding to a brush fire that was threatening structures.

As the apparatus rounded a curve, the right wheels of the apparatus left the roadway and traveled on the shoulder. The driver steered left in an attempt to regain control of the vehicle. The pumper shot across both lanes of traffic, went off the left side of the roadway, came back on the roadway sliding sideways, and rolled 1-1/4-times. The pumper came to rest on the driver's side.

The driver was able to make a radio report of the crash. As other fire departments responded to the scene, the driver and front-seat passenger were removed from the vehicle by passersby. Firefighter Bishop was trapped in the cab with her head lodged between the rear seat and the back wall of the cab on the driver's side. She was removed from the apparatus and transported to the hospital. Firefighter Bishop was pronounced dead upon arrival at the hospital.

The police report indicated that all three firefighters were wearing 3-point seatbelts. State Highway Patrol accident reconstruction experts estimated the speed of the engine prior to loss of control at 74 miles per hour in a 45-mile per hour zone.

The cause of death for Firefighter Bishop was listed as mechanical asphyxia due to blunt head trauma.

For additional information regarding this incident, please refer to NIOSH Fire Fighter Fatality Investigation and Prevention Program report F2002-16 (www.cdc.gov/niosh/face200216.html).

April 10, 2002 - 1135hrs
William Jackson "Jackie" Beard, Jr., Captain
Age 56, Career
Greensboro Fire Department, North Carolina

Captain Beard was conducting live fire training for firefighter recruits in an abandoned apartment building. Over the course of the morning, Captain Beard performed eight evolutions in full structural protective clothing and SCBA.

Captain Beard, again wearing full structural protective clothing and SCBA, ignited the next fire, assured that it grew adequately, and then exited the structure. He removed his SCBA and left it with an air unit to be refilled. He appeared to others on the scene to be tired and not acting normally. He walked to an ambulance standing by at the scene and reported to the paramedic staffing the vehicle that he was not feeling well.

Captain Beard sat down near the ambulance. The paramedic noted his appearance and took vital signs. As the paramedic retrieved medical equipment, Captain Beard slumped over. ALS-level medical care was begun immediately, and Captain Beard was shocked three times with no conversion. CPR was initiated and continued throughout the transport to the hospital. Captain Beard was pronounced dead in the emergency room after efforts there failed to revive him. While he was being treated in the emergency room, it was discovered that the tube that had been inserted to assist in his breathing had been placed in a position where he was not being properly ventilated.

The cause of death was listed as a probable cardiac arrhythmia secondary to ischemic heart disease caused by severe coronary artery arteriosclerosis.

For additional information regarding this incident, please refer to NIOSH Fire Fighter Fatality Investigation and Prevention Program report F2002-19 (www.cdc.gov/niosh/face200219.html).

April 11, 2002 - 1030hrs
Earl M. Hemphill, Fire Chief
Age 61, Career
Russell City Fire Department, Kansas

Chief Hemphill and members of his department were dispatched as mutual aid to a motor vehicle crash. Chief Hemphill and four other firefighters responded to the scene in a rescue truck.

The rescue truck was the first fire department unit on the scene and found an overturned SUV and a camper that had been in tow. The staff of an ambulance that had arrived on the scene first reported that there were no injuries that required the assistance of the fire department. Chief Hemphill and other firefighters turned their attention to the fluids leaking from the SUV.

As Chief Hemphill and another firefighter inspected the front of the SUV, they noticed another piece of fire apparatus approaching the scene at a high rate of speed. The driver of the apparatus was waving his hands back and forth to indicate that he had no brakes. Chief Hemphill ran one way and the other firefighter ran the other way. The apparatus struck Chief Hemphill and threw him down an embankment. The apparatus struck another vehicle, left the roadway, overturned, and ejected the driver.

Firefighters and EMS workers on the scene provided treatment for Chief Hemphill and he was transported to the hospital. Despite extraordinary efforts on the part of the responders on the scene and the staff at the hospital, Chief Hemphill was later pronounced dead at the hospital. The cause of death was listed as chest, abdominal, and skull trauma.

The apparatus involved in the collision had a history of mechanical problems. The driver received minor injuries and was discharged from the hospital the following day. The apparatus driver was initially charged with driving too fast for existing conditions and operating an unsafe vehicle. The charges were later dropped.

For additional information regarding this incident, please refer to NIOSH Fire Fighter Fatality Investigation and Prevention Program report F2002-18 (www.cdc.gov/niosh/face200218.html).

April 13, 2002 - 1230hrs
William J. Tripp, Jr., Firefighter
Age 28, Volunteer
Richford Fire Department, New York

Firefighter Tripp and the members of his department responded to a minor single-vehicle crash. As he laid flares to protect the scene, Firefighter Tripp complained of not feeling well. He was treated by a paramedic on the scene and transported to the hospital. He was pronounced dead at the hospital.

The death was attributed to natural causes. No autopsy was performed, so it is impossible to specify the exact cause of death.

April 27, 2002 - 0830hrs
John A. Nuber, Firefighter
Age 56, Career
Erie Bureau of Fire, Pennsylvania

Firefighter Nuber reported to work at his normal assignment after his regular 3-day off period. Shortly after reporting for duty, he complained of feeling ill. His co-workers recommended that he go to a hospital, but he declined and drove himself home. Just after arriving at home, he started having chest pains. EMS was called and found Firefighter Nuber in cardiac arrest upon their arrival.

Treatment was provided, and Firefighter Nuber was transported to the hospital. He was pronounced dead shortly after arrival at the hospital.

May 3, 2002 - 2115hrs
Derek Duval Martin, Firefighter
Age 38, Career
St. Louis Fire Department, Missouri

Robert "Rob" Bruce Morrison, Firefighter
Age 38, Career
St. Louis Fire Department, Missouri

Firefighter Martin and Firefighter Morrison responded as a part of a rescue company crew to the report of a fire in a 2-story commercial building.

First-arriving firefighters found light smoke showing and a fire on the first floor. While other firefighters opened up the building, engine company firefighters advanced a hoseline into the first floor area and knocked down the fire. As the ceiling on the first floor was pulled, fire was noted in the

APPENDIX

space between the first and second floors. Fire extension into the second floor was suspected. The handline was removed from the first floor and advanced to the second floor.

An engine company captain became separated from his crew at the rear of the first floor of the building. He opened a roll-up door for egress. The fresh air supplied by the open door allowed the remaining fire on the first floor to progress rapidly. A metal security gate at the base of the roll-up door prevented his escape. The captain was able to escape when firefighters and civilians at the rear of the structure moved the gate to permit his exit. While he was trapped, the captain made a number of requests for assistance on the radio.

At the same time, firefighters from the rescue company were opening up the second floor. An engine company firefighter came upon Firefighter Morrison as they worked on the second floor. Firefighter Morrison appeared to be lost and conditions in the area were worsening. The firefighter attempted to lead Firefighter Morrison to the exit but almost became disoriented himself. As he worked his way to the exit, he came upon Firefighter Morrison lying face down and unresponsive. The firefighter was unable to move Firefighter Morrison and, running out of air himself, he was forced to leave the structure. As soon as he exited the building, the firefighter notified a chief officer that Firefighter Morrison was down.

A search party was organized, including Firefighter Martin. The search party entered the building and located Firefighter Morrison. Firefighter Morrison was removed from the building and provided with emergency medical aid.

The captain of the rescue company did another head count and realized that Firefighter Martin was now missing. A second search party entered the building and was aided in the discovery of Firefighter Martin by the sound of his PASS device. Firefighter Martin was removed from the structure and emergency medical care was provided.

Firefighter Morrison was missing for approximately 20 minutes and Firefighter Martin was missing for approximately 29 minutes.

Firefighter Morrison had a blood carboxyhemoglobin level of 47.9 percent and third-degree burns over 18 percent of his body. Firefighter Martin had a carboxyhemoglobin level of less than 10 percent and suffered third-degree burns over 40 percent of his body. Firefighter Martin was pronounced dead upon his arrival at the hospital. Firefighter Morrison died the next day.

Both firefighters were promoted to Captain posthumously.

For additional information regarding this incident, please refer to NIOSH Fire Fighter Fatality Investigation and Prevention Program report F2002-20 (www.cdc.gov/niosh/face200220.html).

May 4, 2002 - 1743hrs
Christopher Nicholas Kangas, Junior Firefighter
Age 14, Volunteer
Brookhaven Fire Company, Pennsylvania

Junior Firefighter Kangas was attending a barbeque when he heard the fire sirens of his local fire station. After stopping at a police station to attempt to learn the nature of the response, he rode his bicycle through an intersection controlled by a stop sign but failed to stop. While in the intersection, he was struck by an automobile.

Junior Firefighter Kangas struck the windshield of the automobile with his head and fell to the ground. Members of Junior Firefighter Kangas' department responded to the scene and provided treatment. He was transported to a local hospital and later transported to a children's hospital. He was pronounced dead early the next day. The cause of death was listed as multiple trauma.

Junior Firefighter Kangas was not wearing a bicycle helmet. He died 2 days before his 15th birthday.

For additional information regarding this incident, please refer to NIOSH Fire Fighter Fatality Investigation and Prevention Program report F2002-21 (www.cdc.gov/niosh/face200221.html).

May 6, 2002 - 1015hrs
Thomas W. Kickler, Firefighter
Age 38, Volunteer
Laurens County Fire Department, South Carolina

Firefighter Kickler remained home ill on May 6, 2002, from his full-time job as Assistant Chief with the Pelham-Batesville Fire Department. He was also a volunteer member of the Laurens County Fire Department.

At 0944, Firefighter Kickler responded with members of his volunteer department to a fire in a doublewide mobile home. Firefighter Kickler participated in interior fire control on the scene. During the mop-up stage of the fire, Firefighter Kickler complained of not feeling well. He went to the IC, reported his condition, and left the scene for home.

That evening, Firefighter Kickler was found dead in his home by a friend. The time of death was set approximately 45 minutes after Firefighter Kickler left the scene of the mobile home fire. The cause of death was a heart attack.

APPENDIX

May 14, 2002 - 1700hrs
Jeremy Brown, Firefighter
Age 27, Volunteer
Screven County Fire Department, Georgia

Firefighter Brown was responding in his personal vehicle to a fire involving a log skidder. He was struck with a heart attack; his vehicle left the roadway, and it crashed into a forested area. He suffered from a congenital heart disease that had given him problems throughout his life.

Firefighter Brown had been scheduled to undergo heart surgery in June.

May 22, 2002 - 0420hrs
Sekou Turner, Firefighter
Age 28, Career
Alameda County Fire Department, California

Firefighter Turner and the members of his crew were treating an elderly heart attack victim. Firefighter Turner was directing an ambulance as it backed out of the scene when he collapsed. Other firefighters on the scene began CPR immediately and Firefighter Turner was transported to the hospital. He was pronounced dead at 0500hrs. The cause of death was believed to be a heart attack.

Alameda County Fire Department Web site -- www.co.alameda.ca.us/fire/

May 27, 2002 - 1551hrs
Rodney Chambers, Firefighter
Age 51, Volunteer
Verona Fire Department, Mississippi

Firefighter Chambers and his son responded to their fire station due to the report of a car fire. The fire turned out to be a backfire in a carburetor and Firefighter Chambers returned home. Within minutes of arriving home, Firefighter Chambers suffered a heart attack.

Members of the Verona Fire Department responded to Firefighter Chambers' house and provided assistance, including the use of an Automatic External Defibrillator (AED). He was transported to the hospital and pronounced dead.

May 27, 2002 - 0510hrs
Terry W. Stinson, Firefighter
Age 44, Part-Time (Paid)
Brown Township Fire-Rescue, Indiana

Firefighter Stinson had just returned from an ambulance run. He and members of his department had transported a patient from a nursing facility to the hospital. The response and transport occurred over approximately 1-1/2 hours.

Firefighter Stinson was completing paperwork associated with the incident when he suffered a heart attack. Another firefighter heard a noise and went into the other room to investigate. Firefighter Stinson was found on the floor. Medical care, including the use of an AED, was started immediately, and he was transported to the hospital.

Firefighter Stinson remained on life support until his death on June 7, 2002.

Brown Township Web site -- www.fire-ems.net/firedept/view/moresville2IN

May 29, 2002 - 0615hrs
Gerald Wayne Allbritton, Engineer
Age 45, Career
Columbus Fire Department, Mississippi

Engineer Allbritton and the members of his crew returned from an EMS call at approximately 0215hrs on May 29, 2002. Engineer Allbritton went to bed. At approximately 0615hrs, a firefighter tried to rouse him and found that he had died during the night.

The cause of death was listed as sudden cardiac death due to hypertensive heart disease and coronary artery disease. Engineer Allbritton had undergone his most recent annual medical screening on July 17, 2001.

May 31, 2002 - 0900hrs
Robert Broussard, Firefighter
Age 59, Career
Sycuan Fire Department, California

Firefighter Broussard was participating in an annual recertification process to be re-certified to fight wildland fires. He was part way through the test when he collapsed. A medic unit on standby at the certification course was at his side within a minute and began treatment. Firefighter Broussard was airlifted to a hospital where he was pronounced dead.

The cause of death was listed as a heart attack.

Sycuan Nation Web page -- www.sycuan.com/home.htm

June 8, 2002 - 1318hrs
Shane Matthew Kelly, Firefighter/EMT
Age 26, Career
Oviedo Fire-Rescue, Florida

Firefighter Kelly was off duty and traveling with his wife on a highway in their personally owned vehicle. A heavy rain was falling. As they drove, they came upon the scene of a crash; a vehicle was on its top in the median. Firefighter Kelly exited his vehicle and began to provide medical care to the victims of the crash, according to medical protocol. This incident occurred outside the area protected by Oviedo Fire-Rescue.

APPENDIX

At least six people, including a medical doctor, stopped to provide assistance. As treatment was being provided, a tractor-trailer was unable to stop for backed-up traffic and drove through the median area. Everyone in the median area was either struck by the truck or the secondary impact between the truck and the car that had been involved in the original crash. As firefighters and EMS workers began to arrive in response to the initial crash, Firefighter Kelly was found in a prone position in the median. He was not breathing and had no pulse.

From the moment he stopped to assist at the crash scene, Firefighter Kelly was considered to be on duty by his department. The fire department responsible for the area where the crashes occurred requested that members of Oviedo Fire-Rescue respond to the scene.

The cause of death was listed as multiple blunt trauma injuries. The medical doctor who stopped to help was also killed. The driver of the tractor-trailer was charged with two counts of vehicular manslaughter and other charges.

June 14, 2002 - 1215hrs
Paul Kelly Jolliff, Private
Age 37, Career
Indianapolis Fire Department, Indiana

Private Jolliff was engaged in dive rescue training to become certified as a rescue diver. This was the last day of a 21-day training curriculum.

He was engaged in an evolution that required him to descend to approximately 60 feet and retrieve a concrete block. The purpose of the exercise was to give the students experience with using tools to retrieve a heavy item in zero visibility and bring that item to the surface. The students used a rope tied to a 3-pound weight on the bottom as an ascent/descent line.

Visibility at the bottom was very poor. At some point, Private Jolliff and his partner were attaching ropes to the block as they had been instructed. Private Jolliff experienced some difficulty and lost contact with his partner. The partner went to the surface to get help. A number of other divers joined the search for Private Jolliff. He was found below the surface entangled in ropes. Private Jolliff was brought to the surface and then brought to shore. He was pronounced dead at the scene.

The cause of death was listed as drowning with barotraumas (gas expansion from the blood on ascent). The Indianapolis Fire Department suffered another rescue diver training death in 2000 when Firefighter Warren Smith died.

Indianapolis Fire Department Web site -- www.indygov.org/ifd/newindex.htm

June 17, 2002 - 1445hrs
Steven Ray Wass, Pilot
Age 42, Wildland Contract
Hawkins and Powers Aviation (Wyoming) under contract to the United States Forest Service in California.

Craig Lebare, Co-Pilot
Age 36, Wildland Contract
Hawkins and Powers Aviation (Wyoming) under contract to the United States Forest Service in California.

Michael Harlow Davis, Flight Engineer
Age 59, Wildland Contract
Hawkins and Powers Aviation (Wyoming) under contract to the United States Forest Service in California.

Pilot Wass, Co-Pilot Lebare, and Flight Engineer Davis were the crew of a C-130A airtanker fighting a fire near Walker, California.

As the aircraft began to make a slurry drop run, the wings of the aircraft separated from the main body of the plane (fuselage) near the wing roots. Fire was ignited in the area of the separated wings and the aircraft crashed. All three crew members were killed.

The aircraft involved in this incident was manufactured by Lockheed in 1957 and placed in service by the Air Force. The C-130 is a 4-engine turboprop aircraft. This aircraft began its civil aviation career in 1988. The wildland firefighting version of the C-130A is capable of delivering 3,000 gallons of firefighting agent. An initial evaluation of the plane's records by the National Transportation Safety Board (NTSB) found that the plane had accumulated over 20,000 flight hours.

The wildland fire eventually consumed over 15,000 acres north of Yosemite National Park.

In December of 2002, the United States Forest Service permanently grounded all C-130A aircraft in its fleet, as well as the PB4Y-2 aircraft. A PB4Y-2 was involved in a fatal crash later in the summer.

NTSB accident number LAX02GA201 can be located through the NTSB aviation accident database at www.ntsb.gov/ntsb/query.asp

APPENDIX

June 21, 2002 - 1715hrs
Bartholomew Bailey, Firefighter
Age 20, Wildland Contract
Grayback Forestry (Oregon) under contract to the United States Forest Service in Colorado.

Jake Martindale, Firefighter
Age 20, Wildland Contract
Grayback Forestry (Oregon) under contract to the United States Forest Service in Colorado.

Daniel Rama, Firefighter/Crew Leader
Age 28, Wildland Contract
Grayback Forestry (Oregon) under contract to the United States Forest Service in Colorado.

Retha Shirley, Firefighter
Age 19, Wildland Contract
Grayback Forestry (Oregon) under contract to the United States Forest Service in Colorado.

Zachary Zigich, Firefighter
Age 18, Wildland Contract
Grayback Forestry (Oregon) under contract to the United States Forest Service in Colorado.

The five firefighters were passengers in a van headed from Oregon to Colorado to fight the Hayman fire. There were a total of 11 firefighters in the van and a total of 8 vans in a caravan.

The caravan had just stopped in Parachute, Colorado for food and fuel. As the vans traveled along a rural interstate highway, the 21-year-old driver of the van containing the five firefighters lost control. The driver reached for a cup and was distracted when the van veered left onto the median and rolled a total of four times. Four of the firefighters died immediately and the fifth died on June 24, 2002, as the result of his injuries.

Only one of the five firefighters that died was wearing a seatbelt at the time of the crash. Four of the six surviving firefighters were wearing seatbelts.

The firefighters were traveling in a Ford Econoline E-350 Super Duty van. The National Highway Traffic Safety Administration (NHTSA) reported that stretch vans such as the one involved in this crash are 3 times more likely to roll over when they carry 10 or more passengers. The NHTSA expressed concerns about vans manufactured by Ford, Chevrolet, GMC, and Dodge.

The driver of the van pleaded guilty to one charge of careless driving and was sentenced to 50 hours of community service.

June 30, 2002 - 1813hrs
Richard Alan Cusson, Firefighter Trainee
Age 30, Volunteer
South Killingly Fire Department, Connecticut

Members of Firefighter Cusson's fire department were on the scene of a mutual-aid fire in a lumber yard. Firefighter Cusson arrived on the scene and told the fire chief that he had not been wearing his pager and that he missed the dispatch. The fire chief directed him to return to the fire station to retrieve his protective clothing and then return to the scene.

Firefighter Cusson was driving his personal vehicle, a 1996 Ford Ranger pickup.

As he drove to the fire station, Firefighter Cusson passed a vehicle that had yielded for his blue light. He then entered a left-hand curve, and his vehicle left the roadway and went onto the right shoulder. Firefighter Cusson steered the vehicle back off the shoulder but was unable to maintain control. Firefighter Cusson's vehicle left the right side of the road and impacted a utility pole and a large-diameter tree. Firefighter Cusson, who was not wearing a seatbelt, was partially ejected from his vehicle.

The fire chief was notified of the crash and instructed the dispatch center to send a mutual-aid fire department to the scene. The fire chief also responded and found Firefighter Cusson dead at the scene.

The police incident report related to this crash estimated Firefighter Cusson's speed to be in excess of 77 miles per hour in a 40 mile per hour zone. The cause of death was listed as blunt trauma of the chest.

Firefighter Cusson's father is the Captain of the Fire Police for the South Killingly Fire Department and his brother is an Assistant Chief.

July 1, 2002 - 0719hrs
Kim Alan Granholm, Captain
Age 28, Volunteer
Thomson Township/Esko Volunteer Fire Department, Minnesota

Captain Granholm and members of his department responded to the scene of a car fire on a local interstate highway. The driver of the vehicle reported that the fire was out but firefighters continued to the scene to make sure. A pumper arrived on the scene and parked downstream from the involved vehicle on the right shoulder. A police cruiser parked upstream of the incident on the right shoulder with its emergency warning lights in operation.

An approaching sedan slowed as it came upon the scene. The sedan was rear-ended by a pickup truck that was towing a trailer. The collision propelled the sedan into a spin and the vehicle veered to the right. The sedan struck the police car, spun into the group of firefighters and others on the scene, and pushed the vehicle that originated the incident into the pumper.

Captain Granholm was struck by one of the involved vehicles. Two other firefighters, a police officer, and a civilian were also struck. Captain Granholm was airlifted to a regional hospital where

APPENDIX

he died in surgery. The cause of death was listed as multiple injuries with laceration of the aorta and exsanguination (blood loss).

Thompson Township/Esko Volunteer Fire Department Web site – www.fire-ems.net/firedept/view/EskoMN

July 2, 2002 - 1745hrs
Alan Wayne Wyatt, Firefighter
Age 51, Wildland Part-Time
United States Forest Service San Juan National Forest, Colorado

Firefighter Wyatt was a seasonal tree faller hired by the Forest Service. He first worked the Million fire and was then directed to report to the Missionary Ridge fire on July 1, 2002. He reported for work the morning of July 2, 2002, and was assigned to assist clearing dangerous trees near a road in an area that had already burned.

After a pause in work because of dangerous winds, fallers were briefed on potential hazards and on the work plan for the afternoon. Firefighter Wyatt began to cut down a tree that had been marked for removal. Out of his line of sight, a fire damaged tree fell in his direction. The tree fell from behind Firefighter Wyatt.

A skid operator who saw the incident from a distance ran to the aid of Firefighter Wyatt. The tree that trapped him was cut into pieces and he was freed. Firefighter Wyatt suffered severe head injuries and a broken leg. He was moved to a safe area by other firefighters where it was determined that Firefighter Wyatt had died. The cause of death was listed as head trauma.

The tree that fell and crushed Firefighter Wyatt was a large aspen. The roots of the tree had been weakened by fire, allowing the tree to fall. Firefighter Wyatt was not working with a swamper (assistant) at the time of the incident. One of the duties of a swamper is to keep watch for hazards that may be out of the faller's view.

July 3, 2002 - 0810hrs
Andrew James Waybright, Recruit Firefighter
Age 23, Career
Frederick County Department of Fire and Rescue Services, Maryland

Recruit Firefighter Waybright was in the third day of a 20-week recruit training course. The temperature was 84 degrees and the heat index was 96 degrees.

The day began at 0700 with physical training which included a 1/2-mile walk, a 3.7-mile run, 15 to 20 minutes of calisthenics, and 2 sets of uphill wind sprints, each covering about 300 feet. No water was provided to the recruits during the physical training. The effects of the exertion on the recruits was not recognized by the instructors who led the recruits during the physical training.

During a jog returning to the training facility, Recruit Firefighter Waybright complained of feeling dizzy and collapsed. He was found to be in cardiac arrest; his skin was cold and clammy. CPR was initiated and paramedics responded to the scene. When Recruit Firefighter Waybright arrived at the hospital, his core temperature was 107.6 degrees. He was pronounced dead at 0921. The cause of death was listed as hyperthermia.

In August of 2000, the recruit academy was criticized for cruel treatment of recruits, including forcing recruits to do push-ups for sips of water. During their investigation of the July, 2002, incident, Maryland Occupation, Safety, and Health (MOSH) cited Frederick County with two "serious" safety violations. The recruit training program was closed in January of 2003.

Recruit Firefighter Waybright was also an EMS Captain with the Harney Volunteer Fire Department in Maryland.

July 4, 2002 - 0136hrs
Thomas G. Stewart III, Firefighter/EMT
Age 30, Career
Gloucester City Fire Department, New Jersey

James E. Sylvester, Fire Chief
Age 31, Volunteer
Mount Ephraim Fire Department, New Jersey

John D. West, Deputy Fire Marshal
Age 40, Career
Camden County Fire Marshals Office, New Jersey

The Gloucester City Fire Department was dispatched to a structure fire in a residential building. While units were responding, dispatch advised units of a working fire with people trapped. When firefighters arrived on the scene, they found a well-involved fire in a 3-story wood-frame structure with fire threatening a connected exposure building of the same size. Heavy smoke was showing from the exposure. Mount Ephraim Fire Department's rescue company was also dispatched upon the report of a working fire.

The amount of fire in the building of origin prohibited an interior attack. The initial arriving officer saw movement in a window on the second floor of the exposure. Firefighters were directed to stretch an attack line into the exposure for search and rescue and fire control. Firefighters found that fire had extended into the second floor of the exposure and conducted suppression efforts but were unable to locate any occupants.

Facing heavy fire in the original structure and extension into the exposure, master streams were applied into the original structure. The bulk of the fire was knocked down in the original fire building while crews continued to operate in the exposure. An occupant (mother) was found in the rear portion of the first floor of the exposure by interior crews and was removed from the structure. An interior crew reported a missing firefighter and the structure was evacuated to conduct an accountability of operating personnel. The firefighter was almost immediately accounted for yet there were still three building occupants (children) that had not been located. A crew of eight firefighters and chief officers, including Firefighter Stewart, Chief Sylvester, and Deputy Fire Marshal West, entered the front of the exposure structure to conduct a search. At the 30-minute mark since the dispatch of the incident, the interior crews reported that they were leaving the structure due to conditions. Within seconds of these reports, both the original fire-involved structure and the exposure structure experienced a catastrophic collapse. The collapse occurred approximately 34 minutes after the initial alarm of fire.

Two firefighters freed themselves after the collapse. Four firefighters were trapped in the collapse. Rescue efforts began immediately and two of the firefighters were freed, with the first taking approximately 25 minutes and the last removed almost 1-1/2 hours after the collapse. After an extensive recovery and rescue effort, the bodies of Firefighter Stewart, Chief Sylvester, and Deputy Fire Marshal West were found and removed from the rubble. Three children, who resided in the original fire structure, were also killed in the incident.

The cause of death for all three firefighters was fixed compression as the result of being crushed by the collapse.

Deputy Fire Marshal West was also a deputy chief with the Mount Ephraim Fire Department.

Gloucester City Fire Department Web site -- www.fire-ems.net/firedept/view/GloucesterNJ

Mount Ephraim Fire Department Web site -- www.mefd.org

July 5, 2002 - 2230hrs
Barry Lee Dockter, Firefighter
Age 44, Volunteer
Anamoose Fire Department, North Dakota

Firefighter Dockter and members of his department were providing the fireworks display for their community centennial celebration. Firefighter Dockter was assigned to load and ignite two 5-inch fireworks mortars.

Approximately 40 minutes into the show, near the end of the show, witnesses heard Firefighter Dockter yell and saw him run towards one of the brush trucks that was standing by at the launch location. Firefighter Dockter appeared to be covered with sparks and was attempting to brush something from his clothing. Seconds later a shell, presumably under his left arm, exploded and Firefighter Dockter fell to the ground.

Other firefighters rushed to his aid and an ambulance was called. CPR was started by EMT's that were on the scene. Firefighter Dockter was transported to the hospital where he was pronounced dead several hours later. The cause of death was listed as blunt force injuries to the chest and left upper extremity.

For additional information regarding this incident, please refer to NIOSH Fire Fighter Fatality Investigation and Prevention Program report F2002-31 (www.cdc.gov/niosh/face200231.html).

July 18, 2002 - 1845hrs
Rick Schwartz, Pilot
Age 57, Wildland Contract
Hawkins and Powers Aviation (Wyoming) under contract to the United States Forest Service in Colorado

Milt Stollak, Pilot
Age 56, Wildland Contract
Hawkins and Powers Aviation (Wyoming) under contract to the United States Forest Service in Colorado

Pilots Schwartz and Stollak were operating Air Tanker 123, an aircraft carrying 2,000 gallons of agent and 550 gallons of fuel. They were in the process of performing a fire-retardant drop on the Big Elk fire near Lyons, Colorado. The plane had flown seven previous air attack missions during the day.

During the approach for an eighth drop, the left wing of the aircraft separated from the fuselage. Fire began as the wing separated, and the aircraft pitched nose down until it crashed into the terrain. Both pilots were killed in the crash.

The aircraft involved in this crash was a World War II era PB4Y-2 Privateer which had originally been used as a military aircraft. After the crash, the Forest Service permanently grounded this type of aircraft.

NTSB accident number DEN02GA074 can be located through the NTSB aviation accident database at www.ntsb.gov/ntsb/query.asp

July 28, 2002 - 0200hrs
Steven K. Oustad, Engine Captain
Age 51, Wildland Full-Time
USDA Forest Service, Klamath National Forest, California

Heather J. DePaolo, Firefighter
Age 29, Wildland Full-Time
USDA Forest Service, Klamath National Forest, California

John Seth Self, Firefighter
Age 19, Wildland Full-Time
USDA Forest Service, Klamath National Forest, California

The Type III engine commanded by Captain Oustad was assigned to provide support during the night shift along a narrow forest road as a part of the Stanza fire near Happy Camp, California. The company was assigned to patrol the road, which was being used as a fire break, to prevent the spread of fire. The task involved driving up and down the road repeatedly through the night. Eight hours of the 12-hour night shift had passed.

As the engine patrolled the road, the driver's side tires left the outside edge of the roadway. The engine rolled off the road and down a steep slope (approximately 1,000 feet) into an old timber harvest area. The apparatus rolled a number of times. Two rear-seat occupants sustained serious injuries but remained in the vehicle. Firefighter DePaolo, the driver, Captain Oustad, the front-seat passenger, and Firefighter Self, a rear-seat passenger, were ejected and killed.

The cause of death for all three firefighters was listed as internal trauma. The investigation concluded that all of the occupants of the engine were wearing seatbelts but that the severity of the roll ejected the three firefighters who were killed. Mechanical failure was ruled out. Other factors which may have contributed to the incident were darkness, smoke, and dust in the air due to passing vehicles.

July 30, 2002 - 1000hrs
John K. Mickel, Lieutenant
Age 32, Career
Osceola County Fire-Rescue, Florida

Dallas Brandon Begg, Firefighter
Age 20, Career
Osceola County Fire-Rescue, Florida

Members of Osceola County Fire-Rescue and the Orlando Fire Department were conducting live fire training in a vacant 1,462 square foot single-story concrete block single-family residential occupancy. The residence was on the site of a college that had been closed.

Prior to the start of the exercises, all firefighters on the scene participated in a walk-through of the structure. The object of the exercise was for two firefighters to enter the structure in full structural protective clothing and SCBA and search for a rescue mannequin dressed in protective clothing. Once the mannequin was located, the firefighters were to remove the mannequin from the structure. Two other teams were assigned to enter the structure and control the fire. Four instructors were placed inside the structure to monitor safety. A Rapid Intervention Team (RIT) crew consisting of two firefighters stood by outside the structure.

During the first evolution of the day, the mannequin was placed in the kitchen in the western portion of the structure. A bedroom in the northeast portion of the structure was used as the point of origin for the fire. The bedroom was just inside of the front door of the structure on the right side of the hallway.

Fuel for the fire consisted of pallets and straw placed in and outside of a closet in the bedroom. The fuel was ignited with a road flare. Some time prior to the beginning of the training evolution, a foam mattress was placed on top of the burning pallets and straw.

Lieutenant Mickel and Firefighter Begg entered the structure as the search and rescue team at 1011. They entered the bedroom to perform a search as the first attack team entered the structure behind them. The second attack team stood by at the front door. When firefighters entered the structure, they found zero visibility with heavy heat and smoke.

The safety officer monitoring the interior lost track of the search and rescue team, thinking that they had exited the fire room and passed him. The safety officer began to search the rest of the structure in an attempt to locate Lieutenant Mickel and Firefighter Begg. At that time, the attack team began to apply water in short bursts to the fire room as the windows of the room were broken out by a firefighter on the exterior. A great deal of steam was produced as the water was applied. When the window was broken out, the fire room flashed over. The firefighter who had ventilated the window reached inside and found the heat damaged shell of a firefighter's helmet.

The helmet discovery was reported to the IC. The IC asked repeatedly for an accountability report from the search and rescue team with no answer. The second attack team entered the structure and extinguished the fire in the area of origin. After the fire was controlled, the firefighters from the second attack team began to overhaul the room. They found a form on the floor in firefighter's protective clothing and assumed that the form was the training mannequin.

The IC did not receive any response to his calls to the search and rescue team, so he ordered a personnel accountability report and ordered the RIT team to enter the structure and find the search and rescue team. The time was 1020. Positive personnel accountability reports were received from all teams except the search and rescue team. The IC ordered an evacuation of the structure. As the second attack team began to leave the structure, they reached down to drag out the form that they thought was a mannequin and found that the form was Lieutenant Mickel. As Lieutenant Mickel was removed through the window, the body of Firefighter Begg was found. The report of firefighters down was transmitted at 1024.

Both firefighters were provided with ALS-level EMS services and transported to the hospital where they were pronounced dead. The cause of death for both firefighters was listed as burns and smoke inhalation.

The carbon monoxide level for Lieutenant Mickel was found to be 23 percent and the level for Firefighter Begg was found to be 22.5 percent.

Firefighter Begg had been employed by Osceola County Fire-Rescue for 8 days. Lieutenant Mickel was a 9-year veteran of the department.

A complete copy of the Florida State Fire Marshal's report on the incident can be downloaded from www.fldfs.com/SFM/bfai/OsceolaDeathReport.htm

July 30, 2002 - 1845hrs
Leonard Gordon Knight, Pilot
Age 52, Wildland Part-Time
Geo-Seis Helicopters (Colorado) under contract to the United States Forest Service in Colorado

Pilot Knight was operating his helicopter fighting the Big Elk fire in the Rocky Mountain National Park. He was the sole occupant of the aircraft. Pilot Knight was dropping water on hot spots along the fire boundary.

Just prior to the crash, witnesses heard Pilot Knight report that his helicopter was going down and saw the helicopter drop to the ground with the blades turning slowly. Pilot Knight was killed on impact. His death was attributed to head injuries.

The 32-year-old French helicopter involved in the crash had a history of mechanical problems spanning the 9 months prior to the crash.

NTSB accident number DEN02GA085 can be located through the NTSB aviation accident database at www.ntsb.gov/ntsb/query.asp

August 1, 2002 - 1445hrs
David Arthur Martin, Firefighter
Age 48, Volunteer
Opal Volunteer Fire Department, South Dakota

Firefighter Martin and the members of his department responded to a wildland fire. He was riding in the back of a pickup operating a hoseline along the boundary of the fire. The pickup was equipped with a slide-in tank/pump unit.

Firefighter Martin was thrown from the pickup and landed in the flames at the head of the approaching fire. The driver of the pickup was not aware that Firefighter Martin had been thrown off. When the driver realized that Firefighter Martin was gone, he attempted to locate him. The driver's view of the scene was obstructed by smoke and flames. When Firefighter Martin fell, he lost his eyeglasses.

After the fall, Firefighter Martin began to walk or run for about 200 feet in an attempt to escape the flames. He attempted to seek refuge from the fire in a brush covered dry creek bed but was blocked by a fence. He then walked or ran an additional 374 feet in an unburned area along a fence. He was discovered by other firefighters and moved to a safe haven. He was transported by helicopter to a regional hospital and then transferred to a burn treatment facility.

Firefighter Martin died as the result of complications from his burns on August 6, 2002. He was burned over 80 percent of his body. The fire was intentionally set and his death was classified as a homicide.

August 12, 2002 - 1611hrs
Roger M. "Mikie" Dunn, Captain
Age 48, Volunteer
Clute Volunteer Fire Department, Texas

Captain Dunn and the members of his department responded to a mutual-aid structural fire involving a single-family residence. A patrolling police officer had discovered the fire and reported heavy visible smoke and fire.

Captain Dunn drove Clute Engine 801 to the scene. Captain Dunn assisted another firefighter with donning his personal protective equipment and SCBA and then walked to another engine to speak with another firefighter. Shortly after he arrived at the engine, he collapsed of an apparent heart attack.

EMS care was initiated immediately, and Captain Dunn was transported to the hospital. He failed to respond to treatment and was pronounced dead at the hospital.

The cause of death was listed as arteriosclerotic cardiovascular disease and hypertensive cardiovascular disease.

Clute Volunteer Fire Department Web site -- www.geocities.com/Clutevfd800/index.html

August 12, 2002 - 1638hrs
Travis Lyn Wiens, Firefighter
Age 28, Volunteer
Wichita West Volunteer Fire Department, Texas

Firefighter Wiens responded as the passenger in a brush engine that responded to a wildland fire. When the engine arrived on the scene, Firefighter Wiens boarded the extended front bumper of the 5-ton converted military cargo truck and began to apply water on the fire. The road on which the truck was operating was obscured by heavy smoke.

As the truck reached the end of the fire front, the driver began to make a U-turn across the roadway to make another pass at the fire front. The truck was in the middle of the U-turn when it was struck by a 3/4-ton pickup truck that drove through the smoke.

The impact caused the driver of the fire truck, who was not wearing a seatbelt, to be thrown across the interior of the cab. The impact also threw Firefighter Wiens off the front platform and knocked his helmet off. Firefighter Wiens landed on the ground with his head in front of the left front wheel of the truck. The truck, which was still in gear with no one behind the wheel, rolled over Firefighter Wiens and inflicted massive head injuries.

Firefighter Wiens was pronounced dead at the scene. Criminally negligent homicide charges were filed against the driver of the fire apparatus and the driver of the pickup.

August 17, 2002 - 0319hrs
William H. Goodrich, Jr., Assistant Chief
Age 56, Volunteer
North Hampton Volunteer Fire Department, Pennsylvania

Chief Goodrich was the first firefighter on the scene of a working attic fire in a house. The fire occurred during a severe thunderstorm.

Chief Goodrich assisted firefighters on the first-arriving engine company as they extended an attack line to the second floor of the structure. After completing the task, Chief Goodrich left the structure and continued in his role as the IC.

The fire was rapidly knocked down by the attack line. The officer that was on the attack line left the building and reported knockdown to Chief Goodrich. As the attack line officer was removing his protective clothing and SCBA, Chief Goodrich collapsed from an apparent heart attack.

EMS treatment was initiated immediately by paramedics who were standing by at the scene. Chief Goodrich was transported to the hospital but was pronounced dead. Chief Goodrich had a history of heart problems and high blood pressure.

APPENDIX

August 30, 2002 - 1945hrs
Harold Coons, Jr., Fire Police Captain
Age 76, Volunteer
South Schodack Fire Department, New York

Fire Police Captain Coons responded to the scene of a serious two-vehicle crash. Upon his arrival on the scene, Fire Police Captain Coons began the task of controlling traffic and protecting the emergency scene.

As a medical helicopter arrived on the scene, Fire Police Captain Coons was seen to drop to one knee and then fall to the ground, the victim of a heart attack. ALS-level EMS care was initiated immediately by medical personnel who had arrived on the helicopter.

Emergency medical care was continued by EMS personnel who had responded to the original crash, and Fire Police Captain Coons was transported to the hospital. It was later reported that Fire Police Captain Coons had been pronounced dead at the hospital.

Fire Police Captain Coons had held his office for 25 years and was the chairman of the board of fire commissioners for his department at the time of his death.

The driver of the vehicle responsible for the original crash was charged with driving while intoxicated and additional charges related to the death of Fire Police Captain Coons.

September 3, 2002 - 2343hrs
Joseph J. Craft, Firefighter
Age 33, Volunteer
Penn Hills Universal Fire Company #6, Pennsylvania

Firefighter Craft had returned earlier in the evening from his first night of Essentials of Firefighting training. Later that night, firefighters were dispatched to the report of a structural fire during an electrical storm.

Firefighter Craft responded as a member of the crew of a support truck. The report of fire ended up to be a malfunctioning garage door opener in a house that had been struck by lightning.

As the truck returned to the station, firefighters in the vehicle saw Firefighter Craft lay his head against the side of the truck and assumed that he was sleeping. Upon their arrival at the station, they discovered that Firefighter Craft was unresponsive.

Paramedics were called to the scene and an AED was applied. Despite these efforts, Firefighter Craft could not be revived and was pronounced dead on September 4, 2002. The cause of death was listed as tetrallogy of fallot with pulmonary atresia, a congenital heart defect.

September 5, 2002 - 1458hrs
Jason Kevin Jackson, Firefighter
Age 17, Volunteer
Almaville Volunteer Fire Department, Tennessee

A small brush fire was reported, and the Almaville Volunteer Fire Department was dispatched to respond. The first firefighter to arrive on the scene reported that the fire had grown and that additional assistance would be required.

Firefighter Jackson had arrived at the fire station and called the chief on the radio to see if the department's tanker truck would be needed. The chief directed Firefighter Jackson to bring the tanker to the scene in a nonemergency (no lights or siren) mode.

Firefighter Jackson was the sole occupant and driver of a 2002 model Freightliner tanker with a 2,000-gallon water tank. As he traveled to the scene, the right wheels of the tanker left the paved roadway. Firefighter Jackson applied the brakes and turned the steering wheel sharply to the left. The back end of the tanker came around and the tanker experienced a rollover and ended up in a field to the right side of the roadway.

The crash was reported to the members of the Almaville Volunteer Fire Department working on the scene of the brush fire. Members of the department responded to the scene and called for additional assistance upon their arrival. Firefighter Jackson was trapped in the cab of the truck and a difficult extrication was required to remove him. Upon his removal, Firefighter Jackson was transported to the hospital where he was later pronounced dead.

The water tank on the tanker was full. Firefighter Jackson had received emergency vehicle operations driver training in 2000.

The cause of death was listed as multiple trauma.

Almaville Volunteer Fire Department Web site -- www.almavillefire.org

September 10, 2002 - 0835hrs
Roger Glen McMillin, Assistant Fire Chief
Age 44, Volunteer
Martin Volunteer Fire Department, South Dakota

Chief McMillin was at work at Mueller Feeds and engaged in the cleaning of a below grade molasses tank used in the manufacturing of animal feed. The tank had a few inches of molasses remaining at the bottom.

A co-worker descended into the tank as Chief McMillin stood by at the tank entrance. In less than one minute, the co-worker became unconscious and fell face down into the molasses. Chief McMillin called for help indicating that he thought that his co-worker had suffered a heart attack. He instructed his supervisor to call 9-1-1 and said that he was going into the tank to assist. From the moment the fire department was called, Chief McMillin was considered to be on duty.

Chief McMillin descended into the tank and was overcome almost immediately. Both men were removed from the tank by arriving responders and CPR was initiated. Both men were transported to the hospital where they were pronounced dead.

The cause of death for Chief McMillin was listed as asphyxiation due to exposure to hydrogen sulfide and a low oxygen atmosphere.

September 14, 2002 - 2255hrs
Michael Raymond Kruse, Firefighter
Age 53, Career
Muscatine Fire Department, Iowa

Firefighter Kruse and members of his department were dispatched to a report of a structure fire in a 3-story multifamily residence. Firefighter Kruse and another firefighter responded in the department's aerial tower. The first firefighters on the scene reported light smoke showing. The first and second floors were clear but firefighters encountered heavy heat and smoke conditions that prevented their entry to the third floor. The order to ventilate the roof was given.

Firefighter Kruse and the other firefighter ascended to the roof of the structure in the platform of the aerial tower. Firefighter Kruse was not wearing an SCBA, the second firefighter was wearing an SCBA. When they arrived at the roof, both firefighters got off the platform. The other firefighter completed the roof cut with a chain saw but did not open up the roof. Due to the smoke conditions on the roof, Firefighter Kruse had been covering his face with his hands.

When the roof cut was complete, Firefighter Kruse pulled on the other firefighter's arm and indicated that they urgently needed to get off the roof. As both firefighters headed for the aerial tower platform, Firefighter Kruse fell to his hands and knees. The other firefighter attempted to grab Firefighter Kruse and lead him to the platform but he was unsuccessful. At this point, Firefighter Kruse turned on his back and fell through the roof into the fire area.

Firefighters in the interior of the structure heard radio transmissions indicating that Firefighter Kruse had fallen through the roof. They fought their way into the third floor of the structure. Firefighter Kruse was located and removed from the fire area. He was then brought outside the structure approximately nine minutes after falling through the roof. He was immediately transported to the hospital.

The cause of death was listed as smoke inhalation. The carboxyhemoglobin level in Firefighter Kruse's blood was 30.3 percent.

September 19, 2002 - 1845hrs
Gerald W. Nadeau, District Fire Chief
Age 51, Career
Fall River Fire Department, Massachusetts

District Chief Nadeau complained of respiratory cold-like symptoms following a structure fire on September 19, 2002. This condition got progressively worse resulting in hospitalization. Upon hospitalization, his condition continued to worsen. He was subsequently diagnosed with adult respiratory distress syndrome which ultimately caused his death on October 24, 2002.

The cause of death was listed as progressive respiratory failure with a clinical history of adult respiratory distress syndrome due to inhalation injuries.

September 23, 2002 - 1103hrs
Cassandra "Sandy" Myers Billings Powell, Firefighter
Age 32, Career
McLeansville Fire Department, North Carolina

Firefighter Powell drove a 1,250-gallon tanker to the department's main station from her substation for repairs to a valve. After the repairs were completed, Firefighter Powell departed the main station for her return trip.

As she drove along a 2-lane road, the right wheels of the apparatus left the right side of the road. Firefighter Powell overcorrected to the left; the back end of the tanker came around to the right, and the apparatus began to slide sideways. The apparatus left the opposite side of the road and came to rest on its roof. Firefighter Powell's head and upper torso were pinned between the steering wheel and the roof.

Firefighter Powell's department was dispatched to the scene and found that they could not gain access to her due to damage to the cab. Partial extrication was completed, and EMS personnel on the scene confirmed that Firefighter Powell had died. The cause of death was listed as a cervical spine fracture and blunt force trauma to the head and neck.

Firefighter Powell was operating the apparatus within the posted speed limit of 55 miles per hour. The apparatus traveled 213 feet from the point where it first left the road to its final resting place. According to the police report on the incident, Firefighter Powell was not wearing a seatbelt at the time of the crash.

Web site for the McLeansville Fire Department -- www.fire-ems.net/firedept/view/McLeansvilleNC

September 30, 2002 - 0210hrs
Ralph Stott, Jr., Captain
Age 50, Career
Terre Haute Fire Department, Indiana

Captain Stott and other members of his department responded to the report of a structure fire in a auto body shop. Captain Stott was in command of an engine company.

The first engine company to arrive reported a working fire with heavy smoke showing. Upon their arrival, Captain Stott and his crew deployed a 1-3/4-inch handline into the front door of the structure. The on-duty battalion chief arrived on the scene and conducted a sizeup of the building. The chief decided to switch to a defensive mode of operations.

The IC attempted to contact Captain Stott and the other firefighter on the handline but was unsuccessful. A lieutenant entered the building and brought Captain Stott and his firefighter to the exterior. The IC ordered Captain Stott and his firefighter to the rear of the building with their handline for exposure protection. As the IC and Captain Stott walked in front of the building, a structural collapse occurred. Captain Stott was buried in the debris.

Firefighters immediately began to dig through the rubble looking for Captain Stott. He was located and removed from the pile. CPR was initiated immediately and Captain Stott was transported to the hospital.

Due to the massive injuries inflicted on Captain Stott when he was crushed by the structural collapse, he was pronounced dead at the hospital. The cause of death was listed as blunt force trauma to the head and chest.

An employee of the body shop was arrested and charged with murder and two counts of arson resulting in serious bodily injury.

Terre Haute Fire Department Web site -- www.fire-ems.net/firedept/view/TerreHaute4IN

October 1, 2002 - 2029hrs
George F. "Bat" Batelli, Sr., Firefighter
Age 55, Volunteer
Garfield Fire Company #1, New Jersey

The Garfield Fire Company #1 was dispatched to a mutual-aid structure fire. Firefighter Batelli and another firefighter followed a responding engine company in Firefighter Batelli's personal vehicle. Firefighter Batelli was driving.

As they responded, Firefighter Batelli told his passenger that he did not feel well and lost consciousness. The firefighter who was the passenger in Firefighter Batelli's vehicle, reached over and put the vehicle in "park" and then guided the vehicle to the curb as it slowed. After the vehicle stopped, the firefighter went into a nearby business and called 9-1-1 for assistance.

Arriving police and EMS personnel started CPR, and Firefighter Batelli was transported to the hospital. He was pronounced dead at 2111. He was the victim of an apparent heart attack.

October 9, 2002 - 0830hrs
Kenneth Wayne Taylor, Captain
Age 49, Career
Madisonville Fire Department, Kentucky

Captain Taylor and the members of his department were dispatched to the report of an apartment fire above a garage. Arriving firefighters found a working fire in the bedroom area of the apartment.

Captain Taylor was the IC. Firefighters advanced a 1-3/4-inch handline and knocked the fire down. Six minutes after his arrival on the scene, Captain Taylor collapsed of an apparent heart attack.

CPR was initiated immediately by firefighters on the scene, EMS workers arrived six minutes after Captain Taylor became ill and took over treatment. Captain Taylor was transported to a hospital, arriving there 31 minutes after he collapsed.

Captain Taylor died on October 11, 2002.

October 11, 2002 - 1601hrs
Robert E. Peterson, Firefighter
Age 57, Volunteer
Bad Axe Area Fire Department, Michigan

Firefighter Peterson and the members of his department responded to a fire involving a wheat stubble field. The fire was eventually contained to 35 acres.

Firefighter Peterson was operating a grass truck. He had just refilled his tank and was driving to a hot spot at the northeast part of the field. After the truck stopped at the hot spot, firefighters prepared the pump at the rear of the truck and called to Firefighter Peterson to slightly relocate the truck. After yelling to him two times, firefighters found Firefighter Peterson unresponsive in the cab from an apparent heart attack.

An ambulance was called and firefighters on the scene began CPR. Treatment and CPR were continued in the ambulance on the way to the hospital but Firefighter Peterson did not survive.

October 20, 2002 - Time Unknown
Rupert Allen "Junior" Fuller, Firefighter
Age 76, Volunteer
Darlington-Gaskin Fire Department, Florida

Firefighter Fuller responded to an EMS call with other members of his department. Shortly after his arrival on the scene, he complained of not feeling well. He was transported to a local hospital and later transferred to a regional hospital. Firefighter Fuller died on October 24, 2002, as the result of a CVA.

November 1, 2002 - 0010hrs
Timothy DiOrio, Lieutenant
Age 36, Volunteer
Maine Fire Company of the Coal Township Fire Department, Pennsylvania

Lieutenant DiOrio and the members of his department were dispatched to a mutual-aid structure fire involving a large old house that was being used for storage. During the fire fight, a portion of the wraparound porch collapsed and pinned Lieutenant DiOrio's leg in the rubble.

Two firefighters came to his aid immediately, one tried to pull Lieutenant DiOrio free by his arms and the other attempted to lift the debris off Lieutenant DiOrio's leg. Moments later, the entire house collapsed, propelling the rescuers away from Lieutenant DiOrio.

The secondary collapse covered Lieutenant DiOrio. It took firefighters hours to control the fire and remove the wreckage that covered Lieutenant DiOrio. He was crushed by a large wooden beam and died of traumatic asphyxiation.

Lieutenant DiOrio was also a Pennsylvania State Trooper based in Selinsgrove.

November 4, 2002 - 2220hrs
Edmund E. Malinski, Firefighter
Age 39, Volunteer
Northern Wayne Fire Company, Pennsylvania

Firefighter Malinski and other members of his department were involved with fighting a fire in a 2-story residence. Firefighter Malinski assisted with the deployment of attack lines into the structure.

About 1-1/2 hours into the incident, Firefighter Malinski was found near a stairwell on the second floor. Firefighters removed Firefighter Malinski from the building and began CPR. He was transported to the hospital where he was pronounced dead about an hour after being found unconscious.

The cause of death was listed as carbon monoxide poisoning as the result of smoke inhalation. Firefighter Malinski was wearing full structural protective clothing, a PASS device, and an SCBA. The SCBA's air supply was depleted and the PASS device was found in the "off" position.

The fire was caused by a basement wood stove.

November 10, 2002 - 2145hrs
Robert Glenn Poore, Firefighter/Dispatcher
Age 71, Volunteer
Briceville Volunteer Fire Department, Tennessee

Firefighter Poore and other members of his department were at their fire station on standby during severe weather. After the storm passed, firefighters left the station to assess damage, search for blocked roadways, and assist those in need.

Firefighter Poore's neighborhood was especially hard hit with a number of homes suffering major damage. As Firefighter Poore assessed the scene, he collapsed of a heart attack. A neighbor discovered him and summoned help. Firefighters responded and began CPR. The response of an ambulance was delayed by fallen trees, so firefighters continued CPR until streets could be cleared enough to allow an ambulance to arrive at the scene. Firefighters and EMS workers continued CPR during the transport to the hospital but Firefighter Poore was not revived.

Firefighter Poore was a founding member of the Briceville Volunteer Fire Department. He had a history of difficulty with his heart.

November 12, 2002 - 1434hrs
Patrick L. Brooks, Firefighter
Age 38, Career
West Hartford Fire Department, Connecticut

Firefighter Brooks was on duty in his fire station. The crew had just finished lunch when Firefighter Brooks said that he was not feeling well and went to lie down. Other firefighters discovered him unresponsive and began medical treatment.

CPR was initiated and an AED was applied. The AED delivered five shocks prior to the arrival of paramedics. Firefighter Brooks was transported to the hospital but efforts to revive him failed.

The cause of death was listed as natural; an inflammation of the heart caused by a disease called sarcoidosis. Firefighter Brooks was the president of the International Association of Fire Fighters Local 1241.

November 18, 2002 - 1630hrs
Bruce S. Fletcher, Fire Chief
Age 50, Volunteer
Ashford Volunteer Fire Department, Connecticut

Chief Fletcher had just returned from an ambulance run. After the other members of the crew left the station, Chief Fletcher began to do the paperwork associated with the run. He began to suffer breathing difficulty and called his wife. His wife, in turn, called 9-1-1. Firefighters from Chief Fletcher's department and other organizations responded to the station.

Upon their arrival, they found Chief Fletcher unresponsive. He had administered oxygen to himself but was now unconscious and without a pulse. CPR was initiated and ALS care was provided with the arrival of paramedics. Chief Fletcher was transported to the hospital but efforts to revive him failed.

Ashford Volunteer Fire Department Web site -- www.ashfordfire.org

November 25, 2002 - 1415hrs
Randall E. Carpenter, Lieutenant
Age 45, Career
Coos Bay Fire & Rescue, Oregon

Jeffrey Edward Common, Firefighter
Age 30, Volunteer
Coos Bay Fire & Rescue, Oregon

Robert Charles "Chuck" Hanners, Firefighter
Age 33, Volunteer
Coos Bay Fire & Rescue, Oregon

Coos Bay Fire & Rescue was dispatched to the report of a fire in a building that contained a truck and auto supply store and a machine shop. The fire was first discovered by building occupants who smelled smoke and discovered warm walls and fire behind the wall in an upstairs bathroom. The occupants attempted to fight the fire prior to calling the fire department. When firefighters arrived on the scene, they reported a light smoke condition.

Firefighters advanced two attack lines into the building. Lieutenant Carpenter, Firefighter Common, and Firefighter Hanners were working on the second floor of the building. Finding no visible fire, the firefighters opened up the walls and ceiling to expose and extinguish the fire. When the fire hidden in these concealed spaces was exposed to fresh air, it progressed rapidly.

Firefighters on the roof reported that the roof was feeling spongy, and the IC ordered an evacuation of the building. At approximately the same time, the roof and other structural supports over the second floor collapsed. Fire spread rapidly throughout the building.

Lieutenant Carpenter and Firefighter Common were trapped under the debris on the second floor. Firefighter Hanners, who may have been descending the stairs at the time of the collapse, was propelled down the stairs by the force of the collapse and ended up behind a customer service counter.

A personnel accountability report found that three firefighters were missing. Other firefighters advanced attack lines into the building to search for the trapped firefighters. Firefighter Hanners was discovered and removed from the building. Due to fire conditions and structural instability, firefighters were unable to reach Lieutenant Carpenter and Firefighter Common until the fire was controlled.

All three firefighters died of asphyxiation and exposure to heat and smoke. Lieutenant Carpenter had a carboxyhemoglobin level of 53 percent and Firefighter Common's level was 49 percent. Firefighter Hanners' carboxyhemoglobin level at autopsy was 3 percent. The medical examiner noted that Firefighter Hanners had undergone extensive life-saving measures including the administration of oxygen, and that the level of carbon monoxide in his blood was likely much higher prior to his removal from the building.

The fire was caused when heat from the flue of a propane-fueled incinerator/parts cleaner ignited structural components in the wall and ceiling of the building's second floor. The fire may have burned in these concealed spaces for as long as 4 hours prior to discovery. The cleaner had been installed without a permit, and it was improperly installed. The owner of the business and the installer were later charged with criminally negligent homicide.

Coos Bay Fire & Rescue Web site -- www.coosbay.org/departments/fire/fire.html

December 4, 2002 - 1845hrs
Kerry Neis, Firefighter
Age 31, Career
Fort Rucker Fire Department, Alabama

Firefighter Neis was participating in an Aircraft Rescue Fire Fighting (ARFF) training exercise. Firefighter Neis was a career civilian military firefighter.

The exercise involved the use of a Military Adapted Commercial Item (MACI) fire truck. The Amertek 2500 is a combined structural and ARFF apparatus with roof and bumper turrets for ARFF operations and preconnected hoselines for structural and ARFF incidents.

The apparatus was placed in pump and idled up for the exercise. Somehow, the truck slipped into road gear and lurched forward. The apparatus struck two firefighters. Firefighter Neis was killed and the other firefighter was injured.

December 5, 2002 - 2010hrs
Michael Lee Depauw, Captain
Age 51, Career
Dallas Fire-Rescue Department, Texas

Captain Depauw and his engine company were among the first to arrive at the scene of an attic fire in a 2-story residence. While fighting the fire from the second floor of the home, Captain Depauw suddenly collapsed. At first, the members of his crew thought that he had tripped over something. They realized that something was wrong when he failed to get up.

Captain Depauw was removed from the building and CPR was initiated. An AED was applied and ALS-level assistance was provided. He was pronounced dead at Medical City Dallas Hospital.

The accidental fire eventually went to two alarms. Captain Depauw's death was caused by a heart attack. He had a history of heart difficulty.

Dallas Fire-Rescue Web site -- www.dallasfiredept.com

APPENDIX

December 5, 2002 - 2130hrs
Dennis Harris, Firefighter
Age 43, Volunteer
Mount Vernon Volunteer Fire Department, Tennessee

Firefighter Harris was operating a pump at a structure fire when he was struck with a heart attack. He was transported to the hospital but was not revived.

December 7, 2002 - 0750hrs
Henry James "HJ" Wissel, Fire Chief
Age 55, Volunteer
Heidelberg Fire Department, Pennsylvania

Chief Wissel was among the first to arrive at a working structural fire in a large automotive and towing company.

Chief Wissel arrived on the first engine company and was operating the pump in support of an aggressive attack on the fire. Two minutes after his arrival on the scene, Chief Wissel collapsed of an apparent heart attack. Chief Wissel was treated on the scene and transported to the hospital, where he died.

The fire eventually went to eight alarms and involved several structures and a number of vehicles.

December 12, 2002 - 1432hrs
Jonathan Myron Lanphear, Firefighter
Age 23, Volunteer
Boyd Volunteer Fire Department, Minnesota

Firefighter Lanphear was in the process of getting a haircut when his pager activated. He and the other members of his department were dispatched to a trash fire. Firefighter Lanphear began his response to the incident in his personal vehicle, a 1999 Grand Am.

Firefighter Lanphear approached an intersection that was controlled by a stop sign at a speed too great to stop. Although he attempted to stop prior to the stop sign, his vehicle began to skid. Firefighter Lanphear's vehicle overran the intersecting road, turned sideways to the left, and descended a steep embankment. The passenger side wheels of the vehicle were knocked off, the vehicle rolled twice, and the vehicle ended up on its roof.

Firefighter Lanphear, who was not wearing a seatbelt, was partially ejected through the sunroof and pinned underneath the vehicle. Arriving rescue personnel pronounced Firefighter Lanphear dead at the scene. Extrication was later completed.

The cause of death was listed as head trauma.

December 15, 2002 - 1652hrs
George A. Walker, Jr., Firefighter
Age 61, Career
Clarksville Fire Department, Indiana

Firefighter Walker and members of his department responded to a residential structure fire caused by heat from a fireplace that had been added to the house. Firefighter Walker assisted with the raising of a ground ladder. After the fire was knocked down, Firefighter Walker was directed to deliver a message to the pump operator.

The pump operator reported to the IC that Firefighter Walker was talking but not making any sense. EMS personnel on-scene were directed to assist Firefighter Walker. After an assessment, the decision was made to transport Firefighter Walker to the hospital.

Emergency room personnel at the hospital determined that Firefighter Walker was having a heart attack. However, Firefighter Walker refused treatment. The deputy chief of the department was sent to the emergency room to talk with Firefighter Walker. After speaking to the deputy chief, Firefighter Walker consented to treatment. During the treatment for the heart attack, Firefighter Walker suffered a CVA.

Firefighter Walker was admitted to the hospital and suffered other CVA's during the following week. On the weekend after he became sick, Firefighter Walker checked himself out of the hospital and went home for the weekend. When he failed to reappear at the hospital, a paramedic/firefighter stopped by Walker's house to check on him.

Firefighter Walker answered the door and then collapsed into the paramedic/firefighter's arms. EMS was called and Firefighter Walker was again transported to the hospital. After being transferred to a long-term care facility, he died on February 1, 2003.

Clarksville Fire Department Web site -- www.town.clarksville.in.us/departments/fire.html

December 27, 2002 - 0400hrs
Terry Wren Carroll, Firefighter
Age 51, Volunteer
Coats - Grove Fire & Rescue, Inc., North Carolina

Firefighter Carroll was on the scene of an EMS incident. He was the driver of the ambulance. He and other responders had just loaded the patient into the ambulance and Firefighter Carroll went around the side of the ambulance to prepare to drive the unit to the hospital.

Prior to his departure from the scene, Firefighter Carroll suffered a heart attack. Members of Firefighter Carroll's department, paramedics, and mutual-aid responders treated Firefighter Carroll at the scene and en route to the hospital.

Firefighter Carroll was revived just prior to his arrival at the hospital. He was treated and then flown to a regional hospital. Firefighter Carroll died on January 7, 2003, due to multisystem organ failure.

Pre-2002 Incidents

November 23, 1982 - Time Unknown
Joseph Michael Tynan, Jr., Firefighter
Age 35, Career
Brookline Fire Department, Massachusetts

Firefighter Tynan was responding on his engine company to an incident. During the response, Firefighter Tynan fell from the vehicle and sustained a severe head injury. Firefighter Tynan was hospitalized until his death on May 8, 2002.

The results of a lawsuit involving the incident that injured Firefighter Tynan had a major impact on the way that fire apparatus are currently designed and manufactured. The four-door fire apparatus is now a standard across the United States.

January 17, 1997 - 1430hrs
James Timothy Smith, Firefighter/EMT
Age 36, Volunteer
Flomaton Volunteer Fire Department, Alabama

Firefighter/EMT Smith responded with other members of his department to a mutual-aid residential structure fire in Century, Florida. Upon his arrival at the scene, Firefighter/EMT Smith was directed to advance an attack line into the residence for fire control.

After he had moved about halfway into the structure, a structural collapse occurred. Parts of the roof and support system fell onto Firefighter/EMT Smith's back. He was transported to a local hospital and later moved to a regional hospital.

Firefighter/EMT Smith went through a year of nonsurgical treatment for his injury. He had his first back surgery in February of 1998 and died of complications from his seventh surgery on October 6, 2001.

November 10, 2001 - Time Unknown
James H. Hanson, Firefighter
Age 62, Volunteer
Chauncey-Dover Volunteer Fire Department, Ohio

Firefighter Hanson had just returned home after a response to a motor vehicle crash. He took off his hat and boots, and collapsed of an apparent heart attack.